NATURAL DISASTER RESEARCH, PREDICTION AND MITIGATION

# CHILEAN EARTHQUAKE OF 2010: RESPONSE AND LESSONS

# NATURAL DISASTER RESEARCH, PREDICTION AND MITIGATION

Additional books in this series can be found on Nova's website under the Series tab.

Additional E-books in this series can be found on Nova's website under the E-book tab.

NATURAL DISASTER RESEARCH, PREDICTION AND MITIGATION

# CHILEAN EARTHQUAKE OF 2010: RESPONSE AND LESSONS


JAMES T. MARKOU

AND

SEAN A. DURKIN

EDITORS



Nova Science Publishers, Inc.

*New York*

For permission to use material from this book please contact us:
Telephone 631-231-7269; Fax 631-231-8175
Web Site: http://www.novapublishers.com



**Library of Congress Cataloging-in-Publication Data**

Chilean Earthquake of 2010 : response and lessons / editors, James T. Markou and Sean A. Durkin.
p. cm.
Includes index.
ISBN 978-1-62100-184-3 (hbk.)
1. Chile Earthquake, Chile, 2010 (February 27) 2. Earthquakes--Chile. 3. Disaster relief--Chile. 4. International relief--Chile. 5. Humanitarian assistance--Chile. I. Markou, James T. II. Durkin, Sean A.
HV600 2010 .C5 C45 2011
363.64'9580983090512--dc23
2011031653

*Published by Nova Science Publishers, Inc. † New York*

# CONTENTS

# PREFACE

On February 27th, 2010, an earthquake of magnitude 8.8 struck off the coast of central Chile. Centered 70 miles northeast of Chile's second-largest city, Concepcion, at a depth of 22 miles, the earthquake was the second largest ever recorded in Chile and the fifth largest recorded worldwide since 1900. Over 100 aftershocks of magnitude 5.0 or greater were recorded following the initial earthquake. The earthquake and subsequent tsunami, which struck Chile's coast roughly 20 minutes after the earthquake and moved 2,000 feet onto shore in some places, devastated parts of the country. This book analyzes the response to the Chilean earthquake and the lessons learned.

Chapter 1- On February 27, 2010, an earthquake of magnitude 8.8 struck off the coast of central Chile. Centered 70 miles northeast of Chile's second-largest city, Concepción, at a depth of 22 miles, the earthquake was the second largest ever recorded in Chile and the fifth largest recorded worldwide since 1900. Over 100 aftershocks of magnitude 5.0 or greater were recorded following the initial earthquake. The earthquake and subsequent tsunami, which struck Chile's coast roughly 20 minutes after the earthquake and moved 2,000 feet onto shore in some places, devastated parts of the country. Although there are reports of varying casualty numbers, according to Chile's Ministry of the Interior, the official death toll is 507 (497 bodies have been identified; 10 remain unidentified). The numbers of missing persons are not yet known. Approximately 200,000 homes have been badly damaged or destroyed. Estimates suggest as many as 2 million people may have been affected by the earthquake, an unknown number of whom were injured or displaced.

Chapter 2- In July 2010, in an effort to reduce future catastrophic natural disaster losses for California, the American Red Cross coordinated and sent a delegation of 20 multidisciplinary experts on earthquake response and recovery to Chile. The primary goal was to understand how the Chilean society and relevant organizations responded to the magnitude 8.8 Maule earthquake that struck the region on February 27, 2010, as well as how an application of these lessons could better prepare California communities, response partners and state emergency partners for a comparable situation. Similarities in building codes, socioeconomic conditions, and broad extent of the strong shaking make the Chilean earthquake a very close analog to the impact of future great earthquakes on California. To withstand and recover from natural and human-caused disasters, it is essential for citizens and communities to work together to anticipate threats, limit effects, and rapidly restore functionality after a crisis.

In: Chilean Earthquake of 2010 ISBN: 978-1-62100-184-3
Editors: J. T. Markou and S. A. Durkin 

*Chapter 1*

# CHILE EARTHQUAKE: U.S. AND INTERNATIONAL RESPONSE*

***June S. Beittel and Rhoda Margesson***

## SUMMARY

On February 27, 2010, an earthquake of magnitude 8.8 struck off the coast of central Chile. Centered 70 miles northeast of Chile's second-largest city, Concepción, at a depth of 22 miles, the earthquake was the second largest ever recorded in Chile and the fifth largest recorded worldwide since 1900. Over 100 aftershocks of magnitude 5.0 or greater were recorded following the initial earthquake. The earthquake and subsequent tsunami, which struck Chile's coast roughly 20 minutes after the earthquake and moved 2,000 feet onto shore in some places, devastated parts of the country. Although there are reports of varying casualty numbers, according to Chile's Ministry of the Interior, the official death toll is 507 (497 bodies have been identified; 10 remain unidentified). The numbers of missing persons are not yet known. Approximately 200,000 homes have been badly damaged or destroyed. Estimates suggest as many as 2 million people may have been affected by the earthquake, an unknown number of whom were injured or displaced.

The Chilean government, through the Chilean National Emergency Office, is leading the relief operation and coordinating assistance. Despite

---

* This is an edited, reformatted and augmented version of a Congressional Research Service publication, CRS Report for Congress R41112, from www.crs.gov, dated March 11, 2010.

offers of assistance, thus far the international humanitarian relief operation has been limited pending further requests for assistance from the government. In addition, there are more than 16,000 Chilean military personnel providing humanitarian relief and maintaining public order. At least two elements of the Chilean government's initial response have been criticized in Chile. The first is that the coastal and island communities did not receive timely warning about the tsunami waves that caused so many of the reported casualties. The second Chilean government response that has been widely questioned was the speed with which the Chilean military was deployed to quell looting and violence in the disaster zone. While critics point to weaknesses in the initial response, later assessments by disaster managers gave the Chilean government's response higher marks. Many credited the government of President Michelle Bachelet with success given the scope of the disaster and some labeled the government's response a model. The Chilean government's response to the earthquake has been complicated by the fact that President Bachelet left office on March 11, 2010, and has been succeeded by Sebastián Piñera, the leader of a center-right coalition that won the country's recent presidential election. President Bachelet held meetings with President-elect Piñera in order to ease the transition.

On February 27, 2010, President Barack Obama announced that the U.S. government would assist with earthquake rescue and recovery efforts, pending a request from the Chilean Government. On February 28, 2010, U.S. Ambassador to Chile Paul E. Simons issued a disaster declaration, and through the U.S. Agency for International Development (USAID) Office of Foreign Disaster Assistance (OFDA), authorized $50,000 for the initial implementation of an emergency response program. OFDA deployed a 16-member USAID Disaster Assistance Response Team. As of March 10, 2010, USAID/OFDA reports that it has provided $10.7 million for emergency response activities in Chile. The U.S. Department of Defense is also providing limited assistance. Policy issues of potential interest include levels of U.S. assistance to Chile, burdensharing and donor fatigue, tsunamis and early warning systems, and managing risk through building codes. Related legislation includes S.Res. 431, H.Res. 1144, H.R. 4783.

## Current Conditions

On February 27, 2010, an earthquake of magnitude 8.8 struck off the coast of central Chile. Centered 70 miles northeast of Chile's second-largest city, Concepción, at a depth of 22 miles, the earthquake was the second-largest ever recorded in Chile and the fifth-largest recorded worldwide since 1900.[1] Over 100 aftershocks of magnitude 5.0 or greater were recorded following the initial earthquake.[2]

The earthquake and subsequent tsunami, which struck Chile's coast roughly 20 minutes after the earthquake and moved 2,000 feet onto shore in some places, have devastated portions of the country.[3] Preliminary reports indicate that 2 million people have been affected out of a population of 16.6 million, with some 200,000 homes badly damaged or destroyed and more than 500 people confirmed dead. The Chilean government has declared six regions catastrophe zones: Valparaiso, Metropolitana, Libertador O'Higgins, Araucania, Biobío, and Maule. The Maule region was by far the hardest hit, however, accounting for most of the dead, and the humanitarian situation there remains the most pressing.

Infrastructure across the country has been affected with roads destroyed, bridges and power lines down, and some ports forced to close. Early estimates suggest the economic damage could be between $15 and $30 billion U.S. dollars, the equivalent of 10% to 20% of Chile's gross domestic product (GDP). The infrastructure damage could also slow Chile's economy in the near term, jeopardizing the country's nascent recovery from the global financial crisis. Many believe the economic and human toll would have been much worse if not for Chile's stringent building codes and past experience with earthquakes.[4]

The Chilean government is leading the relief operation and coordinating assistance. Despite offers of assistance, thus far the international humanitarian relief operation has been limited pending further requests for assistance from the government. In addition, there are more than 16,000 Chilean military personnel providing humanitarian relief and maintaining public order.

## Preliminary Numbers at a Glance

Although there are reports of varying casualty numbers, according to the Ministry of the Interior, the official death toll is 507 (497 bodies have been identified; 10 remain unidentified). The numbers of missing persons are not yet known. According to the Chilean government, approximately 200,000 houses have been badly damaged or destroyed. Estimates suggest as many as 2 million people may have been affected by the earthquake, an unknown number of whom were injured or displaced.

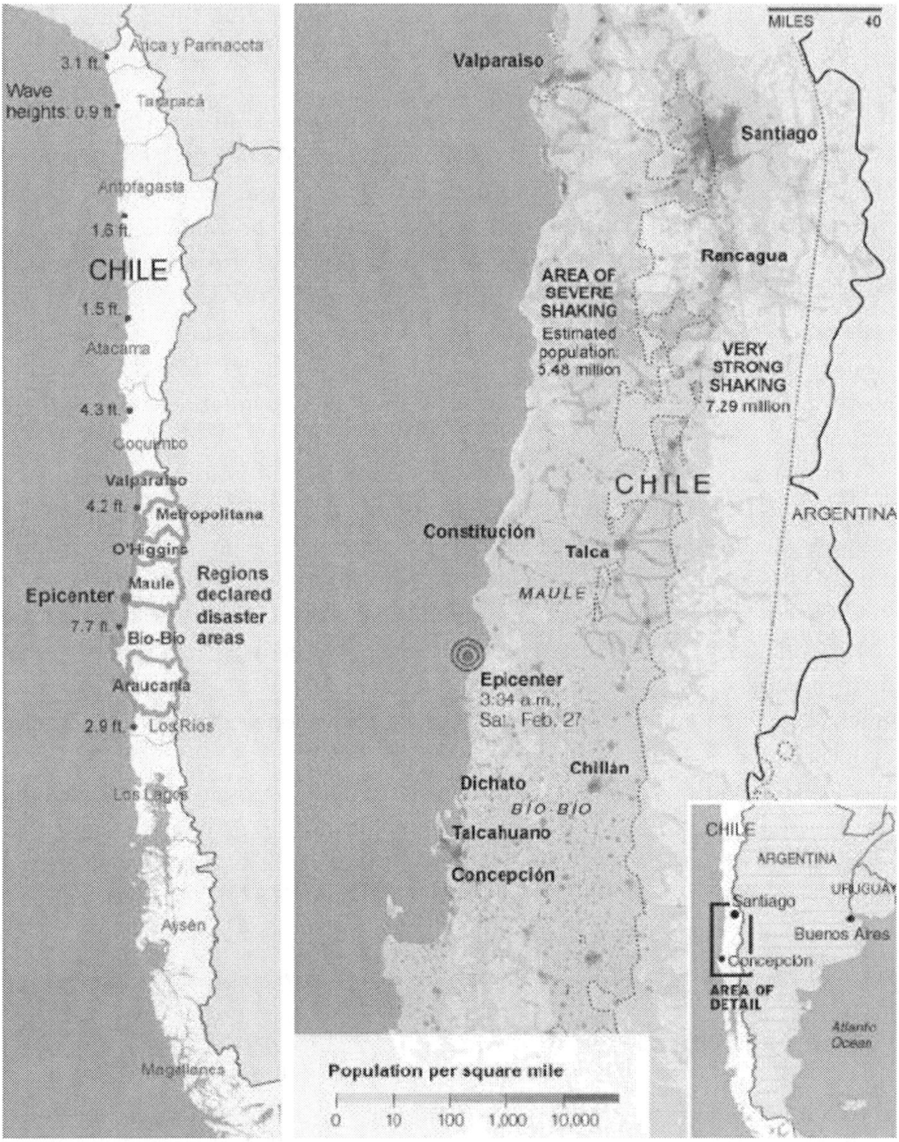


Source: ReliefWeb: UN Office for the Coordination of Humanitarian Affairs (OCHA).

Figure 1. Chile: Regions Affected by Earthquake.

# Chilean Government Response

Since the earthquake, the Chilean government has rushed to provide relief to the victims. In addition to dispatching search and rescue teams and undertaking preliminary needs assessments, the government has worked to reestablish basic services and has distributed food, blankets, and medical equipment in the affected regions. The Chilean National Emergency Office (ONEMI) is coordinating the relief effort. After its initial assessments, the Chilean government appealed to the international community for aid. Among other items, the Chilean government has requested field hospitals, electric generators, satellite communications equipment, water purification systems, and mobile bridges. Chilean officials have met with United Nations (U.N.) agencies to establish a clear plan for international assistance, and the United States and more than 20 other nations have begun to provide Chile with the aid it has requested.[5] (See Appendix B.) Many aid agencies have offered assistance, pending further instructions from the government.

President Michelle Bachelet called the consequences of the earthquake "an emergency without parallel." She issued a State of Catastrophe declaration for the regions of Maule and Biobío on February 28, 2010. The declaration, issued following increasing reports of riots and looting, restricts civil rights and liberties and places the regions under military control to maintain order and avoid social conflict. This is the first time that a State of Catastrophe measure has been employed by the Chilean government since its return to democracy in 1990.

President Bachelet deployed approximately 14,000 military troops to the two regions to restore order based on outbreaks of looting and violence. The military commanders of the regions, in consultation with the Minister of the Interior and the Minister of Defense, declared a curfew for the city of Concepción. The curfew was extended to three cities in the Maule region—Talca, Cauquenes, and Constitución—on March 1. Enforcement of order has led to at least 160 arrests and one death.[6]

At least two elements of the Chilean government's initial response have been criticized in Chile. The first is that the coastal and island communities did not receive timely warning about the tsunami waves that caused so many of the reported casualties. The first tsunami waves hit the coastal city of Valparaiso (330 kilometers northeast of the epicenter) just 34 minutes after the initial earthquake, according to the National Oceanic and Atmospheric Administration (NOAA).[7] But some claim that the tsunami warnings in coastal areas were not clearly transmitted. The Chilean Navy acknowledged their response was partly at fault and a quicker warning would have saved many lives. According to one report, the Chilean Navy received word of the imminent tsunami threatening the coast just 11 minutes after the initial earthquake. However, no warning was issued locally. The President deflected the criticism saying the time is not right to assign blame. On the morning of the catastrophic quake, the President was unable to communicate with many areas in the country. However, after acknowledging their warning system was faulty, the Navy fired the head of its catastrophic warning unit.[8] Bachelet's successor, Sebastián Piñera has called for a thorough "modernization" of the tsunami warning system as his government comes into office.[9]

The second Chilean government response that has been widely questioned is the speed with which the Chilean military was deployed to quell looting and violence in the disaster zone. Some critics have argued that President Bachelet moved too slowly, suggesting she was mistrustful of the military because of her imprisonment and torture under General Augusto Pinochet's military dictatorship more than two decades ago. She began to deploy troops within 48 hours of the earthquake and increased their numbers during the week to a total of 14,000. Troops were welcomed by Chileans both for restoring order and assisting with the ongoing disaster response.

The rural poor were the most affected by the earthquake. In some small cities, food was unattainable when poorer customers were priced out of the market because merchants sharply raised prices. Some reports said that services were restored and aid provided first to middle class areas.[10] The urban poor living near Santiago lost their electricity and water for much longer than those living in the capital which recovered quickly.[11] One of the most vulnerable populations in Chile is the Mapuche, Chile's largest indigenous group comprising approximately 4% of the Chile's 16.6 million citizens.[12] The Mapuche live mainly in Chile's central and southern regions with their seat of power in La Araucanía. They suffered many of the difficulties after the earthquake as other poor Chileans; water and electricity were cut off and houses were destroyed or seriously damaged.[13]

While critics point to weaknesses in the initial government response, later assessments by disaster managers gave the Chilean government higher marks. Aspects of the response that received special attention included: a careful assessment to target foreign assistance so it would complement Chile's rescue and relief efforts; quick restoration of major roads such as the critical North-South highway; and having the military handle logistics. Within ten days of the earthquake, 90% of homes in the disaster area had regular power and water, 500,000 survivors were getting water trucked in, and relief has been distributed by military helicopter, navy ships, and tractor trailers. Many credited the Bachelet government with success given the scope of the disaster and some labeled it a model.[14] In a post quake poll, President Bachelet received an 84% approval rating, leaving office with the highest level of popular support of her presidency.[15]

The Chilean government's response to the earthquake has been complicated by the fact that President Bachelet departed office on March 11, 2010, and was succeeded by Sebastián Piñera, the leader of a center-right coalition that won the country's recent presidential election. Piñera was expected to replace all of Bachelet's cabinet ministers and political appointees, including many of those now overseeing relief operations. Recognizing the need for continuity, Piñera had earlier indicated that he would retain the head of ONEMI and would ask some of the subregional authorities of the six regions declared catastrophe zones to stay on temporarily.[16] On March 9, 2010, President-elect Piñera named key regulators and regional authorities (or intendants as they are called in Chile) to his government before his inauguration, including a new deputy Interior Minister.[17] On March 10, Carmen Fernandez, Director of ONEMI in the Bachelet government, resigned citing weaknesses in the agency's response. President Bachelet held a series of meetings with President-elect Piñera in order to ease the transition.

## Political Transition: March 11, 2010

Governments and non-governmental organizations (NGOs) are conferring with President Bachelet and incoming President Piñera on the nation's needs to smooth the transition in power. President-elect Piñera has declared his government "will be the government of the reconstruction, not the government of the earthquake."[18] He faces huge challenges in a recovery that has been estimated will take 3-4 years and will cost between $15 and $30 billion dollars by the outgoing administration. Piñera has toured the hardest hit regions, and continued to make appointments. He is re-orienting his strategy from promising to return Chile to an annual growth rate of 6% to rebuilding and reconstruction. He has also pledged to improve coordination between authorities and the armed forces; to repair damaged infrastructure such as ports, roads, airports and telecommunication and power lines; and to construct homes in the affected zones.[19] A billionaire, conservative politician and one of Chile's wealthiest investors, will face political challenges such as keeping expectations in check, and keeping right wing elements within his coalition with ties to the Pinochet era under control, in addition to steering Chile out of the disaster.

Chile declared three days of national mourning for victims of the quake from March 7 to March 9, 2010. The tone of the ceremony to inaugurate the new president was austere out of respect for the national tragedy and widespread mourning resulting from the earthquake. Public activities and ceremonies for the President-elect have been cancelled although official events—including visits from foreign delegations—will take place. Both leaders anticipate a smooth transition. A countrywide telethon held in Chile on March 7 (with the participation of President Bachelet and President-elect Piñera and visiting U.N. Secretary-General Ban Ki-Moon) was expected to raise $30 million and brought in almost $60 million. The funds will go to a local NGO "Un Techo Para Chile" (A Roof For Chile) that plans to use the funds for the construction of 23,000 small wooden shelters in the affected areas.

# Status of the Relief Effort

As mentioned above, the Chilean government is leading the relief operation and coordinating assistance. Assessments are ongoing. It remains to be seen how the needs of those affected by the tsunami may differ from those primarily affected by the earthquake and, further, how such differences might be integrated into the overall humanitarian response.

## Humanitarian Needs and Response

### *Health*

Of 130 existing health clinics in the affected areas, six hospitals were evacuated, nine were severely damaged, and 115 remain fully operational. Based on these figures, approximately 4,000 beds (or 20 to 25%) have been lost. Regional health services have now returned to normal levels. Psychological and trauma assistance is needed.

Ten field hospitals and a Red Cross mobile unit have been deployed to affected areas. Chilean military personnel are providing medical assistance. On March 1, the government set up a National Committee of Mental Health in Emergencies and Disasters. The Ministry of Health has provided fortified milk, medicines, and vaccines. It has also announced a general vaccine campaign in high risk areas. The Chilean Red Cross and other Red Cross societies are providing medical assistance and conducting assessments for water supplies and relief items.

### *Food*

Many of the affected areas are without regular access to food. The numbers of those needing food assistance have not yet been determined. In addition, the military is distributing food and reportedly some NGOs are also providing food assistance.

### *Emergency Shelter and Non-Food Items*

The Chilean government reports there are an estimated 90 shelter camps with 19,000 people. A further 25,000 people are in temporary shelters (such as schools) and another 50,000 are in makeshift camps. Many others have reportedly staying with family and friends, and in the case of migrant workers, have returned to their homes in other parts of the country. An estimated 200,000 houses have been destroyed or badly damaged. Two regions – Maule and Biobio – were hit particularly hard by the earthquake and tsunami and in some places 72% of houses were destroyed and 24% require significant repair. In terms of planning for longer-term emergency shelter, the rainy season does not usually start in Chile until May or June.

The International Federation of the Red Cross/Red Crescent (IFRC) has provided relief items to 5,000 families. Volunteers have constructed 300 emergency shelters. The $60 million collected during a weekend telethon in Chile will support the construction of 23,000 emergency shelters. The U.S. Agency for International Development (USAID) has distributed plastic sheeting for 1,500 family shelters. Assessments of shelter needs are ongoing.

### *Water and Sanitation*

More than 150 rural water supply systems, affecting approximately 200,000 people, were damaged. Others are not working due to damage to electrical systems. Access to potable water has generally increased in affected areas since the earthquake. However, in Concepción, where 55% of inhabitants have access to potable water and in Arauco where 50% have access, bottled water is urgently needed. In addition, the Chilean military is providing bottled water in water tankers. Three airplane loads of potable water have also been sent from Bolivia.

### *Education*

It is estimated that up to 33% of the schools in the six areas most affected by the earthquake are either not operational at all or only partially operational. In areas less affected, such as Santiago, classes were suspended until March 8 when approximately 80% of the children returned to school. Classes in areas more affected by the earthquake are likely to start mid March or later.

### *Early Recovery: Emergency Telecommunications and Infrastructure*

It is reported that mobile and land telecommunications have improved significantly in affected areas. In the city of Concepción services are up to 65-70%.

The Chilean Minister of Public Health has said that rehabilitation of roads and other infrastructure will cost $1.2 billion. The Chilean government has identified some structures for rehabilitation. The government of Sweden has offered eight portable bridges. Various international structural engineers are conducting assessments. Electricity and natural gas conduits are being restored to affected areas.

### *Early Recovery: Fisheries and Agriculture*

The government reports that more than 26,000 fisherman lost their homes and basis for livelihood (boats and equipment) as a result of the tsunami. The government is providing assistance for coastal towns and has promised agricultural assistance. Some wine-producing areas have also been affected.

# INTERNATIONAL HUMANITARIAN RESPONSE AND FUNDING

## The Role of the United Nations and Other Organizations

The United Nations has a country team in Santiago, which is lead by the Economic Commission for Latin America and the Caribbean (ECLAC). The U.N. country team includes agencies such the U.N. Development Program (UNDP), the U.N. Children's Fund (UNICEF), the Office of the High

Commissioner for Human Rights (OHCHR), the Pan-American Health Organization (PAHO), the U.N. Educational, Scientific, and Cultural Organization (UNESCO), the Food and Agriculture Organization (FAO), and the International Labor Organization (ILO), as well as other entities such as the European Union and International Organization for Migration (IOM).

The Chilean government did not make a specific request to the U.N. Office for the Coordination of Humanitarian Assistance (OCHA) until days after the initial earthquake. At that point, the government made targeted requests for specific assistance such as communication equipment, water purification apparatus, and field hospitals, which were met largely through bilateral support. OCHA did send a small assessment team to Santiago and Concepción. OCHA remains on stand-by should assistance be requested by the Chilean government. A revised request may be forthcoming from the Chilean government with the March 11 transition of power.

## Other Humanitarian Actors

International recovery efforts are typically complex because they require coordination among numerous different actors. Apart from U.N. agencies, those responding to humanitarian crises include international organizations, NGOs, private voluntary agencies (PVOs), and bilateral and multilateral donors. A great deal of assistance is provided by other governments and international entities.

The International Federation of the Red Cross and Red Crescent Societies (IFRC) is working with the Chilean Red Cross Society and other national red cross societies to provide assistance to earthquake survivors.

Various international NGOs have also mobilized to respond to the crisis.

## International Funding

Many international actors have offered relief assistance to Chile, either through pledges of financial support to the government of Chile or aid organizations or by directly offering relief supplies and emergency personnel.

### *Donor Contributions and Pledges*

The U.N.'s Central Emergency Response Fund (CERF) has pledged $10 million.[20] The U.N. Country Team in Chile is drafting proposals for use of the

CERF funds which are focused on health, water and sanitation, emergency shelter, education and emergency telecommunications.

So far, through governments and the private sector, the international community has pledged millions of dollars in aid, materials, and technical support. Appendix B highlights donor contributions and in-kind pledges. At least 29 countries from around the world have contributed to the relief effort. Obtaining an up-to-date record of all international contributions is not possible—in part because some assistance is not reported to governments or coordinating agencies—and in part because of the delay in their recording.

### *Private Contributions*

Private sector assistance has been given at a much slower rate than private contributions following the earthquake in Haiti. Some reports attribute the slower pace to the fact that Chile is much better prepared to deal with the destruction; others point to the fact that the damage is much less severe. In addition, the Chilean government requested assistance from the United Nations but not so far from international humanitarian groups. Aid groups remain on stand-by pending requests for assistance. Moreover, reports also indicate that Chileans are donating and volunteering in large numbers to fill gaps in the earthquake response. As stated above a weekend telethon organized by a local organization raised $60 million.[21]

### *Response of International Financial Institutions*

Rebuilding in the aftermath of the massive earthquake will take years and require an investment in the reconstruction of as much as $30 billion dollars, or between 15-20% of Chile's current GDP. After an initial assessment, President Bachelet stated that Chile will need loans from the World Bank and other multilateral sources to pay for recovery.[22] The World Bank, the Inter-American Development Bank (IDB) and the International Monetary Fund (IMF) all offered assistance to Chile immediately following the earthquake. The World Bank and the IDB have opened credit lines for Chile.

As one of the region's strongest economies and democracies, Chile is likely to receive the loans it seeks. The country's growth projections for 2010 are estimated at 5 to 5.5% (J.P. Morgan actually bumped up its prediction of Chile's growth rate to 5.5% following the earthquake in anticipation of economic activity during a vigorous reconstruction effort).[23] Chile's credibility in international markets, strong macroeconomic policies, and open markets will be seen favorably.[24] In addition, Chile has reserved more than $11 billion in a sovereign fund, mostly during President Bachelet's administration, earned from copper exports.[25] These emergency reserves are likely to be used to help finance the nation's recovery.

## U.S. HUMANITARIAN ASSISTANCE

Selected activities of key U.S. government agencies—USAID and the Department of Defense (DOD)—are described briefly below.

### USAID

On February 27, President Barack Obama announced that the U.S. government would assist with earthquake rescue and recovery efforts, pending a request from the Chilean Government. USAID set up a Washington, DC-based Response Management Team (RMT) through the Office of Foreign Disaster Assistance (OFDA) to coordinate and facilitate the humanitarian response to the earthquake. On February 28, 2010, U.S. Ambassador to Chile Paul E. Simons issued a disaster declaration, and through OFDA, authorized $50,000 for the initial implementation of an emergency response program. OFDA deployed a 16-member USAID Disaster Assistance Response Team (DART). The RMT is supporting the USAID/DART.

To date, the overall focus of the U.S. government's response has been conducting needs assessments, working with the U.S. embassy in Chile and the Chilean government on priority humanitarian needs, coordinating delivery of USAID/OFDA-funded relief supplies, and providing communications support. See Appendix C for further information about the U.S. Government humanitarian response mechanism.

On March 2, Secretary of State Hillary Clinton met with Chilean President Michelle Bachelet and President-elect Sebastián Pinĕra in Santiago as part of a pre-planned regional trip to Latin America. Secretary of State Clinton delivered 20 satellite phones provided by USAID/OFDA and DOD to assist with communications in areas affected by the earthquake. An additional 40 phones have since been provided.

As of March 10, 2010, USAID/OFDA reports that it has provided $10.7 million for emergency response activities in Chile. USAID/OFDA issues regular situation reports assessing the progress of relief operations.[26]

## Department of Defense[27]

DOD contributions to the earthquake relief efforts in Chile are on a much more limited scale than Operation Unified Response in Haiti. As with Haiti, the U.S. Southern Command (SOUTHCOM) is overseeing DOD's efforts with a command and control team based in Santiago. The U.S. Air Force has deployed two C-130 cargo aircraft and 50 support personnel to provide airlift to affected areas. The Air Force has also deployed an Emergency Medical Expeditionary Support Team with 84 personnel to the city of Angol. The U.S. Navy has dispatched an engineering assessment team to the Chilean naval base Talcahuano to survey and assess damages to the harbor facilities.

# Regional Response

Offers of assistance by countries in the region came quickly. On March 1, President Lula of Brazil was the first foreign leader to visit Chile, and at a meeting with President Bachelet pledged all possible Brazilian assistance. Brazil has sent helicopters and C-130 planes carrying a field hospital and staff. Damage to Chile's power supply was widespread. Because both of Chile's main oil refineries were damaged, President Bachelet pledged to increase gasoline, diesel and fuel oil imports. Neighboring Argentina quickly doubled its normal supply of natural gas to Chile to 5.5 million cubic meters per day on March 3.[28] Argentina also sent an Air Force hospital, food, water, medical supplies, electricity power generators and four water purification plants. Many Latin American countries sent in-kind donations with leaders of Ecuador and Bolivia explaining they were cash-strapped. For example, in the first week following the earthquake Venezuela sent seven tons of food and Bolivia sent 40 tons of drinking water. Cuba provided a 26-member medical team and field hospital. (For more details, see Appendix B).

These quick offers of aid are significant in light of Chile's history of sometimes mistrustful relations with its neighbors due to unresolved border disputes. At the summit of the Union of South American Nations (a regional cooperation body) held in the aftermath of the Haitian earthquake in early February, Latin American leaders agreed to set up a $300 million fund for Haiti's reconstruction and to assist one another in future natural disasters.[29] The response by Latin American countries to Chile's earthquake and tsunami seems to be evidence of this commitment.

## BACKGROUND ON CHILE AND U.S.-CHILE RELATIONS[30]

Following a violent coup against democratically elected Marxist President Salvador Allende in 1973, Chile experienced 17 years of military rule under General Augusto Pinochet before reestablishing democratic rule in 1990. A center-left coalition of parties known as the Concertación has governed Chile over the two decades since the end of the dictatorship. In addition to addressing human rights violations from the Pinochet era, the coalition has enacted a number of constitutional changes designed to strengthen civilian democracy. Chile made significant economic progress under the Concertación's free market economic policies and moderate social programs, and produced notable economic growth and considerable reductions in poverty.

On January 17, 2010, Sebastián Piñera of the center-right Alianza coalition was elected president in a second-round runoff vote, defeating former President Eduardo Frei (1994-2000) of the Concertación. Piñera's election was the first for the Chilean right since 1958, and his inauguration on March 11, 2010, will bring an end to 20 years of Concertación governance. Throughout his campaign, Piñera pledged to largely maintain Chile's social safety net while implementing policies designed to boost the country's economic growth. Even before the earthquake, Piñera was expected to work closely with Concertación to enact his policy agenda because his coalition does not enjoy absolute majorities in either house of Congress.

Chile has enjoyed close relations with the United States since its transition back to democracy. Both countries have emphasized similar priorities in the region, designed to strengthen democracy, improve human rights, and encourage trade. Chile and the United States have also maintained strong commercial ties, which have become more extensive since a bilateral free trade agreement between them entered into force in 2004. Additionally, U.S. officials have expressed appreciation for Chile's leadership and moderating influence in a region increasingly characterized by political unrest and anti-American populism.

## Regular U.S. Bilateral Assistance to Chile

Because Chile is an advanced developing country the small amount of current assistance the country receives is focused on modernizing the Chilean military to improve its capacity to act as a peacekeeping force and to conduct joint operations with the U.S. military. U.S. assistance also goes to programs that deter weapons of mass destruction, improve civilian control over the military, and upgrade military equipment. Chile received $1.2 million in U.S. assistance in FY2009, an estimated $1.7 million in FY2010, and would receive $2.2 million in FY2011 under the Obama Administration's request. Table 1 shows the distribution between three foreign assistance accounts for Chile.

**Table 1. FY2009-FY2011 Funding for Chile by Aid Account ($ in thousands)**

| Account | FY2009 (Actual) | FY2010 (Estimate) | FY2011 (Request) |
|---|---|---|---|
| FMF | 400 | 400 | 750 |
| IMET | 525 | 900 | 960 |
| NADR | 300 | 450 | 500 |
| Total | 1,225 | 1,750 | 2,210 |

Source: U.S. Department of State, FY2011 Congressional Budget Justification for Foreign Operations.

Notes: FMF=Foreign Military Financing; IMET=International Military Education and Training; NADR=Nonproliferation, Anti-terrorism, Demining and Related Programs.

# Policy Issues

Many Members have already expressed a strong desire to support Chile and provide it with necessary assistance.

## Budget Priorities and U.S. Assistance to Chile

Humanitarian assistance generally receives strong bipartisan congressional support and the United States is typically a leader and major contributor to relief efforts in humanitarian disasters.[31] When disasters require immediate emergency relief, the Administration may fund pledges by depleting its disaster accounts intended for worldwide use throughout a fiscal year. The United States has so far provided $10.7 million in immediate aid for Chile. That aid is drawn from existing funds. The international community is also making donations toward meeting immediate needs.

Finding the resources to sustain U.S. aid pledges may be difficult, particularly amid efforts to tackle rising budget deficits by, among other measures, slowing or reducing discretionary spending. After the 2004 tsunami disaster, some Members of Congress publicly expressed concern that funding for tsunami relief and reconstruction, which depleted most worldwide disaster contingency accounts, could jeopardize resources for subsequent international disasters or for other aid priorities from which tsunami emergency aid had been transferred. These accounts were fully restored through supplemental appropriations. At the time, others noted the substantial size of U.S. private donations for tsunami victims and argued that because of other budget pressures, the United States government did not need to transfer additional aid beyond what was already pledged.

In Haiti, the full extent and cost of the disaster is not yet known. Disaster accounts are currently being drawn down to provide relief to Haiti. The State Department reports that in order to respond to future humanitarian crises, these resources would need to be replenished by June 1, 2010. If not replenished, U.S. capacity to respond to other emergencies could be curtailed. Congress will also likely consider a major request to help fund Haiti's recovery and reconstruction. Should substantial funds be required for Chile as well, it is possible this would be added to the request for Haiti.

## Burdensharing and Donor Fatigue

The earthquake in Chile appears to require far less assistance from the international community than in Haiti. It is not always evident whether figures listing donor amounts represent pledges of support or more specific obligations.[32] Pledges made by governments do not necessarily result in actual contributions. It also cannot be assumed that the funds committed to relief actually represent new contributions, since the money may previously have been designated to provide support for the affected country. It will take time to obtain a more complete picture of the actual costs of the disasters in Haiti and Chile and how they will be shared among international donors.

Comparing U.S. and international aid is also difficult because of the often dramatically different forms the assistance takes (in-kind contributions vs. cash, for instance). As the situation in Haiti stabilizes, and attention turns to early recovery and reconstruction, sustaining donor interest in Haiti (and commitment to honor existing pledges) could be a challenge. Moreover, this challenge is compounded by the need to maintain funding priorities and secure funds needed for other disaster areas, such as the recent earthquake in Chile.

## Tsunamis and Early Warning[33]

Due to the fact that the earthquake in Chile occurred offshore, it generated a tsunami which struck parts of the Chilean coastline and offshore islands, causing damage and fatalities. Tsunami warnings were issued by the National Weather Service Pacific Tsunami Warning Center for Hawaii, Japan, and other regions bordering the Pacific Ocean that may have been vulnerable to a damaging tsunami, although most regions far from the epicenter did not experience any serious damage. A tsunami caused significant damage to the city of Hilo, Hawaii, following the May 1960 magnitude 9.5 earthquake that also occurred along the subduction zone fault about 143 miles south of the February 27, 2010, earthquake. Why the 1960 earthquake generated a tsunami that caused damage and fatalities in Hawaii, Japan, and the Philippines while the 2010 earthquake did not is not yet well understood and is being actively studied. In addition, these studies will likely consider the effectiveness of existing early warning systems.

Subduction zone megathrust faults generate the largest earthquakes in the world. The Cascadia Subduction Zone megathrust that stretches from mid-Vancouver island in southern British Columbia southward to Cape Mendocino in northern California has the potential to generate a very large earthquake, similar in magnitude to the February 27 Chilean earthquake. The fault's proximity to the northwestern United States coastline—approximately 50-100 miles offshore— also poses a significant tsunami hazard; destructive waves from a large earthquake along the fault could reach the coast of Oregon and Washington in less than an hour, possibly as soon as tens of minutes.

The Cascadia Subduction Zone fault forms the boundary between the subducting Juan de Fuca tectonic plate and the overriding North American plate, very similar to the relationship between the Nazca Plate and the South American Plate off the Chilean coast. If the Cascadia Subduction Zone megathrust were to "unzip" or rupture along a large section of its entire length, it would likely generate a megathrust earthquake near magnitude 9 or more, similar to the 1964 Alaskan earthquake, the 1960 and 2010 Chilean earthquakes, and the 2004 Indonesian earthquake. Scientists have documented that the last time this occurred along the Cascadia Subduction Zone fault was in 1700. The 1700 earthquake spawned a tsunami that traveled across the Pacific Ocean and struck Japan. Because of the tectonic similarities between the Cascadia Subduction Zone megathrust and the Nazca-South American plate megathrust, scientists hope to learn a great deal about the seismic hazard in the Pacific Northwest by studying the unique strong ground motion recordings from the 2010 Chilean magnitude 8.8 earthquake.

## Managing Risk through Building Codes[34]

The potential damage to Chile's buildings and infrastructure in urban areas was mitigated by the adoption and enforcement of advanced building codes; these codes include requirements aimed at reducing damage from earthquakes' shaking. Assessments of the damage in Chile will provide valuable information on whether changes to Chile's building codes or their enforcement are in order for better managing seismic and tsunami risks. The National Science Foundation and other U.S. federal agencies are supporting or participating in various engineering teams to collect and analyze data from the February 27 event that could inform improvements to international and domestic model building codes.[35]

Chile has demonstrated its willingness and capacity to create, implement, and enforce building codes and other hazard mitigation measures to reduce loss and injury. Chile has the mechanisms and technical capacity to learn from the quake to improve its codes and policies to support more resilient structures and communities (e.g., improving tsunami warning and communication technologies and protocols). Construction that complies with location appropriate building codes can make structures more resistant to collapse and damage; however, structures are never earthquake proof, and increased structural resiliency typically increases construction costs.

Although the Chilean quake was of a larger magnitude than the January 2010 Haitian quake, Chile's urban centers of Concepcion and Santiago experienced less severe shaking than the violent shaking in Port-au-Prince.[36] A combination of more intense shaking, weak building codes and enforcement, and limited preparedness and response capabilities contributed to Haitians facing significantly higher risk than Chileans. How the United States and other nations provide aid to Chile and Haiti is likely to reflect the understanding that multiple actions can reduce risks, including adoption of building codes, in-country technical capacity building, expert technical assistance, and sharing of preparedness and response knowledge and technology.

# APPENDIX A. EARTHQUAKE ZONES

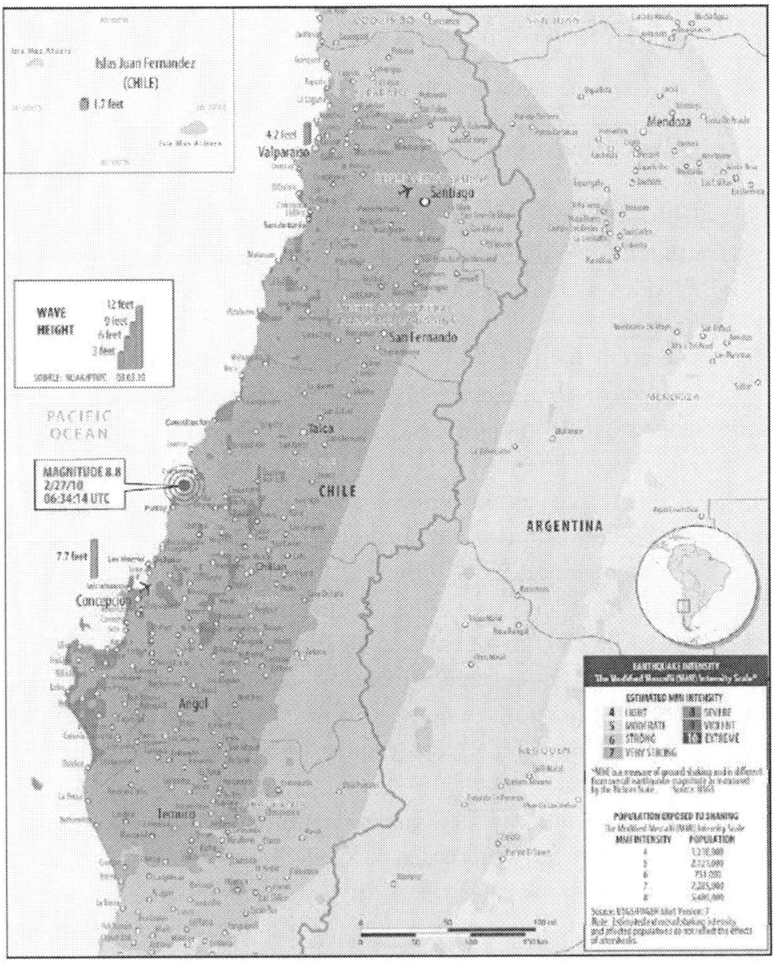


Source: USAID.

Figure A-1. Chile: Earthquake Intensity Zones.

# Appendix B. Donor Contributions and Pledges to Chile in Response to the February 27, 2010, Earthquake

### Table B-1. Donor Contributions and Pledges to Chile (as of March 10, 2010)

| Country/Agency Donor | Monetary Pledge (USD) | In-kind Support Pledge |
|---|---|---|
| Argentina | | delivery of 1800 tons of food, a half million liters of water, four water purification plants and four electricity power generators, Argentine Air Force hospital composed of 12 modules, ten thousand doses of vaccine against hepatitis A |
| Australia | 4,413,063 | fifty generators, 150 family tents and 1,060 collapsible beds |
| Bolivia | | 40 tons drinking water |
| Brazil | | Two C-130s from Brazilian Air Force transporting 18 tons of equipment for a field hospital, and 66 military staff |
| Canada | | provided 150 tents with capacity for 10 people each. |
| China | 2,000,000 | chartered plane with quake-relief: carrying 700 tents, 10,000 blankets, 100 portable generators and two water purifiers, for a total of 96 tons |
| Cuba | | 26-member medical team and field hospital |
| Finland | 539,811 | |
| France | | chartered a special plane to bring to Chile a scientific, technical and humanitarian assistance mission, (including a team of seismological experts and a team of infrastructure experts), plus one ton of seismological material (27 seismological stations, solar panels;, GPS), a complete satellite equipment and 10 Iridium telephones, a dialysis and water filtration machine, first aid medicines, community tents, and five water purification stations |
| Germany | | 4-member rescue team |
| Greece | 269,906 | |
| Indonesia | 1,000,000 | |

**Table B-1. (Continued)**

| Country/Agency Donor | Monetary Pledge (USD) | In-kind Support Pledge |
|---|---|---|
| Japan | 3,336,625 | dispatch of international emergency medical team, electric generators, tents, water purifiers |
| Korea, Republic of | | dispatch of a plane with medical equipments and non-perishable food |
| Malta | | pills for the treatment of 17,000 liters of drinking water and further relief goods |
| Mexico | | 3 tons of water purification equipment (chlorine, etc.) and a team of 8 specialists to evaluate structures |
| New Zealand | 343,407 | |
| Norway | 1,675,042 | |
| Peru | | 2 airplanes with 30 tons of relief items including tents, plastic sheets, water purification unit; search and rescue team of 35 specialists; one field hospital with surgical capacity and in-patient hospital care; 21 doctors |
| Russia | | one plane with 28 MT of relief items (100 tents, 10 portable power generators, 2,000 blankets and 7 tons of food) |
| Singapore | 50,000 | |
| Spain | | 7.5 tons of relief items, one field hospital, and a team of 75 experts (firefighters and sanitation experts) |
| Sweden | 720,041 | |
| Switzerland | 276,753 | dispatched a 3-person regional Rapid Response Team |
| United Kingdom | 382,263 | 600 tents for up to 3,000 people |
| United States of America | a | a |
| Uruguay | | two water purification machines |
| Venezuela | | 7 tons of relief supplies and an assessment team of 27 experts |
| Vietnam | 150,000 | |
| Other: | | |
| Organization of American States | 25,000 | 20 satellite telephones |

| Country/Agency Donor | Monetary Pledge (USD) | In-kind Support Pledge |
|---|---|---|
| Red Cross/Red Crescent | 946,939 | |
| Private (individuals & organizations) | 13,828,311 | |

Source: OCHA at http://www.reliefweb.int/fts. Prepared by Julissa Gomez-Granger, Information Research Specialist, Knowledge Services Group.
[a] For U.S. assistance, please see section on "U.S. Humanitarian Assistance."

## APPENDIX C. THE U.S. GOVERNMENT EMERGENCY RESPONSE MECHANISM FOR INTERNATIONAL DISASTERS

The United States is generally a leader and major contributor to relief efforts in response to humanitarian disasters.[37] The President has broad authority to provide emergency assistance for foreign disasters and the U.S. government provides disaster assistance through several U.S. agencies. The very nature of humanitarian disasters—the need to respond quickly in order to save lives and provide relief—has resulted in a rather unrestricted definition of what this type of assistance consists of at both a policy and an operational level. While humanitarian assistance is assumed to provide for urgent food, shelter, and medical needs, the agencies within the U.S. government providing this support typically expand or contract the definition in response to circumstances. Funds may be used for U.S. agencies to deliver services or to provide grants to international organizations (IOs), international governmental and non-governmental organizations (NGOs), and private or religious voluntary organizations (PVOs). The U.S. Agency for International Development (USAID) is the U.S. government agency charged with coordinating U.S. government and private sector assistance. It also coordinates with international organizations, the governments of countries suffering disasters, and other governments.

The Office of Foreign Disaster Assistance (OFDA) in USAID's Bureau for Democracy, Conflict and Humanitarian Assistance (DCHA) provides immediate relief materials and personnel, many of whom are already abroad on mission. It is responsible for providing non-food humanitarian assistance and can quickly assemble Disaster Assistance Response Teams (DARTs) to assess conditions. OFDA has wide authority to borrow funds, equipment, and personnel from other parts of USAID and other federal agencies. USAID has two other offices that administer U.S. humanitarian aid: Food For Peace (FFP) and the Office of Transition Initiatives (OTI). USAID administers emergency food aid under FFP (Title II of P.L. 480) and provides relief and development food aid that does not have to be repaid. OTI provides post-disaster transition assistance, which includes mainly short-term peace and democratization projects with some attention to humanitarian elements but not emergency relief.

The Department of Defense (DOD) Overseas Humanitarian, Disaster and Civic Aid (OHDACA) funds three Dodd humanitarian programs: the Humanitarian Assistance Program (HAP), Humanitarian Mine Action (HMA) Program, and Foreign Disaster Relief and Emergency Response (FDR/ER). OHDACA provides humanitarian support to stabilize emergency situations and deals with a range of tasks including providing food, shelter and supplies, and medical evacuations. In addition the President has the authority to draw down defense equipment and direct military personnel to respond to disasters. The President may also use the Denton program to provide space-available transportation on military aircraft and ships to private donors who wish to transport humanitarian goods and equipment in response to a disaster.[38]

Generally, OFDA provides emergency assistance for 30 to 90 days after a disaster. The same is true for Department of Defense humanitarian assistance. After the initial emergency is over, assistance is provided through other channels, such as the regular country development programs of USAID.

The State Department also administers programs for humanitarian relief with a focus on refugees and the displaced. The Emergency Refugee and Migration Account (ERMA) is a contingency fund that provides wide latitude to the President in responding to refugee emergencies. Assistance to address emergencies lasting more than a year comes out of the regular Migration and Refugee Account (MRA) through the Population, Migration and Refugees (PRM) bureau. PRM assists refugees worldwide, conflict victims, and populations of concern to the United Nations High Commissioner for Refugees (UNHCR), often extended to include internally displaced people (IDPs). Humanitarian assistance includes a range of services from basic needs to community services.

# APPENDIX D. LEGISLATION IN THE 111TH CONGRESS

S.Res. 431. A resolution expressing profound concern, deepest sympathies, and solidarity on behalf of the people of the United States to the people of Chile following the massive earthquake. Introduced March 3, 2010 and agreed to in the Senate March 4, 2010.

H.Res. 1144. A resolution expressing condolences to the families of victims of the February 27, 2010, earthquake in Chile, as well as solidarity and support for the people of Chile as they plan for recovery and reconstruction. Introduced on March 9, 2010, and agreed to in the House on March 10, 2010.

H.R. 4783. The Act to accelerate the income tax benefits for charitable cash contributions for relief of victims of the earthquake in Chile, and to extend the period from which such contributions for the relief of victims of the earthquake in Haiti may be accelerated. Contributors may deduct their donations made by April 15, 2010 from their 2009 tax return. Introduced in the House March 9, 2010, and passed in the House and introduced in the Senate on March 10, 2010.

# APPENDIX E. HOW TO SEARCH FOR OR REPORT ON AMERICAN CITIZENS IN CHILE AND HOW TO OBTAIN SECURITY INFORMATION UPDATES

This section prepared by Anne Leland, Information Research Specialist, Knowledge Services Group.

The Department of State has provided phone numbers and an e-mail address for reporting on American citizens who have been living or travelling in Chile:

The Chile Task Force phone number is 1-888-407-4747 and its e-mail address is ChileEarthquake@state.gov.

In order to expedite your request, provide the individuals full name, birth date, birthplace, location and contact information in Chile as well as any special circumstances.

American citizens living or travelling in Chile may register via the Department of State website at https://travelregistration.state.gov/ibrs/ui/, or may register in person at the U.S. Embassy in Santiago, located at Avenida Andres Bello 2800, Las Condes. The embassy phone number is 56- 2-330 30 00, the fax number is 56-2-330 30 05. Their after-hours emergency phone number is 56- 2 330-3000, and email address is SantiagoAmcit@state.gov.

Information about making phone calls from Chile to the United States and from the United States to Chile may be found at http://chile.usembassy.gov/calling-us-chile.html.

The Department of State frequently updates security conditions using the following phone numbers: 1-888-407-4747 (from inside the United States) and 1-202-501-4444 (from outside the United States).

## APPENDIX F. LINKS FOR FURTHER INFORMATION

This section prepared by Anne Leland and Julissa Gomez-Granger, Information Research Specialist, Knowledge Services Group

**U.S. Government Agencies**

**U.S. Agency for International Development (USAID)**
http://www.usaid.gov/helpchile/
**U.S. Department of State**
http://www.state.gov/chilequake
***U.S. Department of State Travel Warnings and How to Locate or Report Status of American Citizens***
http://travel.state.gov/travel/cis_pa_tw/pa/pa_4737.html
***U.S. Department of State Embassy, Santiago Chile***
http://chile.usembassy.gov/2010press0304-relief.html

**U.S. Geological Survey**
http://earthquake.usgs.gov/earthquakes/recenteqsww/Quakes/us2010tsa6.php#details

**White House**
http://www.whitehouse.gov/blog/2010/02/27/information-chilean-earthquake-and-tsunamiwarnings

**Other Resources**

**Center for International Disaster Information (CIDI)**
http://www.cidi.org/incident/chile-10b/
**Embassy of Chile, Washington DC**
http://www.chile-usa.org/
**European Commission for Humanitarian Aid (ECHO)**
http:/ec.europa.eu/echo/index_en.htm

**InterAction**
http://www.interaction.org/crisis-list/interaction-members-respond-earthquake-chile
**Inter American Development Bank**
http://www.iadb.org/countries/home.cfm?id_country=CH
**International Monetary Fund**
http://www.imf.org/external/country/chl/index.htm

**Organization of American States/Pan American Development Foundation**
http://www.panamericanrelief.org/ht/d/sp/i/19624/pid/19624

**Pan American Health Organization/World Health Organization**
http://new.paho.org/disasters/index.php?option=com_content&task=view&id=1133&Itemid=103 6

**Red Cross Movement**

***American Red Cross***
http://www.redcross.org/portal/site/en/menuitem.1a019a978f421296e81ec89e43181aa0/?vgnextoid=d0206aafe5517210VgnVCM10000089f0870aRCRD
***Chilean Red Cross***
http://www.cruzroja.cl/
***International Committee of the Red Cross***
http://www.icrc.org/Web/Eng/siteeng0.nsf/htmlall/chile?OpenDocument
***International Federation of the Red Cross***
http://www.ifrc.org/where/country/cn6.asp?countryid=46

**Relief Web**
http://www.reliefweb.int/rw/dbc.nsf/doc104?OpenForm&rc=2&cc=chl

**United Nations**

***United Nations Children's Fund (UNICEF)***
http://www.unicef.org/infobycountry/chile_52861.html
***United Nations News Center***
http://www.un.org/apps/news/story.asp?NewsID=34043&Cr=chile&Cr1=
***United Nations World Food Program (WFP)***
http://www.wfp.org/content/immediate-response-earthquake-chile

**World Bank**

http://web.worldbank.org/WBSITE/EXTERNAL/COUNTRIES/LACEXT/CHILEEXTN/0,,menuPK:325279~pagePK:141159~piPK:141110~theSitePK:325273,00.html

## End Notes

[1] In 1960, southern Chile was struck by a magnitude 9.5 earthquake. Jose Luis Saavedra, "Massive earthquake hits Chile, 214 dead," *Reuters*, February 27, 2010.

[2] For more information on earthquakes see, CRS Report RL33861, *Earthquakes: Risk, Detection, Warning, and Research*, by Peter Folger, and U.S. Agency for International Development, "Chile – Earthquake," March 1, 2010.

[3] "Chile earthquake death toll tops 700," *Latin News Daily*, March 1, 2010. "Maps of the Chile Earthquake," *New York Times*, March 1, 2010.

[4] "Chile: Earthquake could boost medium-term growth," *Oxford Analytica*, March 1, 2010.

[5] United Nations Office for the Coordination of Humanitarian Affairs, "Chile Earthquake: Situation Report #2," March 1, 2010; Gobierno de Chile, "Situación terremoto zona centro sur (Actualiza reporte)," March 1, 2010; Hillary Rodham Clinton, "Remarks With Chilean President Michelle Bachelet," *U.S. Department of State*, March 2, 2010.

[6] "Bachelet decreta primer Estado de Catástrofe desde terremoto de 1985," *El Mercurio* (Chile), March 1, 2010; "Amplían toque de queda en zonas más afectadas por terremoto en Chile," *Agence France Presse*, March 1, 2010; "160 detained, one killed during curfew in quake-struck Chile," *Agence France Presse*, March 1, 2010; "Troops in Chile quake areas to number 14,000: Bachelet," *Agence France Presse*, March 2, 2010.

[7] See NOAA web page on tsunami research at http://nctr.pmel.noaa.gov/chile20100227/.

[8] Terry Wade and Fabian Cambero, "Some Chile quake survivors wait for aid a week on," http://www.alternet.org. March 6, 2010.

[9] Fred Tasker and Marcela Ramos, "Chileans didn't get warnings of arriving tsunami; The Chilean government is being criticized for not alerting people of a tsunami threat after a powerful earthquake one week ago.," *Miami Herald*, March 6, 2010.

[10] Dom Phillips, "Poor are the last to get aid amid class war in quake's ghost city; chile," *The Times*, March 4, 2010.

[11] Daniel Hernandez, "Quake exposes social disparities in Santiago, Chile," *Los Angeles Times*, March 4, 2010.

[12] For more on the Mapuche, see CRS Report R40126, *Chile: Political and Economic Conditions and U.S. Relations*, by Peter J. Meyer.

[13] "La Araucanía Comienza a Ponerse de Pie," from the CONADI website: http://www.conadi.cl/, March 7, 2010.

[14] Michael Warren, "Chile's Bachelet, despite criticism, gets high praise from disaster experts for quake response," *Associated Press Newswire*, March 10, 2010.

[15] The poll was conducted between March 3-March 6 by the firm Adimark and results were released on March 9. See Irenea Renuncio Mateos, "New Chilean President Sworn In, Faces Reconstruction Challenges," *IHS Global Insight Daily Analysis*, March 11, 2010.

[16] In Chile, regional governors are appointed by the president. "Chile: Earthquake could boost medium-term growth," *Oxford Analytica*, March 1, 2010.

[17] Chile Pinera Deputy Interior Min In Office Ahead of Inauguration, *Dow Jones International News*, March 10, 2010.

[18] Daniel Hernandez, "Taking reins, is not office yet; Pinera asserts his authority, though he won't be president of Chile until next week," March 5, 2010.

[19] "Bachelet estimates 3-4 years for reconstruction; Piñera unveils principal strategies," *Business News Americas*, March 4, 2010.

[20] As part of the United Nations' reform process, in March 2006, the CERF was launched based on several earlier resolutions approved by the U.N. General Assembly to strengthen the United Nations' capacity to respond to natural disasters and humanitarian emergencies. It is managed by the Emergency Relief Coordinator and head of UNOCHA.

[21] Andy Markowitz, "Chileans Give in Record Numbers to Fill Quake-Relief Gaps," *The Chronicle of Philanthropy*, March 11, 2010.

[22] Terry Wade and Mario Naranjo, "Chile wants loans, scours for quake survivors," http://www.alertnet.org, March 4, 2010.

[23] Samuel Logan, "Chile After the Quake," ISN Security Watch, March 5, 2010. Accessed at: http://www.isn.ethz.ch/isn/Current-Affairs/Security-Watch/Detail/?lng=en&id=113384.

[24] Irenea Renuncio Mateos, "New Chilean President Sworn In, Faces Reconstruction Challenges," *IHS Global Insight Daily Analysis*, March 11, 2010.

[25] "Counting the Costs," *Economist*, March 6, 2010.

[26] See USAID website: http://www.usaid.gov/our_work/humanitarian_assistance/disaster_assistance/countries/chile /template/index.html.

[27] Prepared by Steven Bowman, CRS Specialist in National Security, Foreign Affairs, Defense, and Trade Division.

[28] "Argentina doubles natgas flow to quake-hit Chile," *Reuters-AlterNet*, March 3, 2010.

[29] "Chile 'quake prompts Argentine help," *Latin News Daily*, March 9, 2010; Christian Völkel, "UNASUR to Rally Around in Support of Haiti's Reconstruction," *IHS Global Insight Daily Analysis*, February 10, 2010.

[30] This summary is drawn from CRS Report R40126, *Chile: Political and Economic Conditions and U.S. Relations*, by Peter J. Meyer.

[31] For background information see CRS Report RL33769, *International Crises and Disasters: U.S. Humanitarian Assistance, Budget Trends, and Issues for Congress*, by Rhoda Margesson.

[32] Relief Web is a good source of information. See http://www.reliefweb.int.

[33] Prepared by Peter Folger, CRS Specialist in Energy and Natural Resources Policy, Resources, Science, and Industry Division.

[34] Prepared by Nicole Carter, CRS Specialist in Natural Resources Policy, Resources, Science, and Industry Division.

[35] Since the mid-1990s, Chile has adopted parts of the U.S. model building code. Assessment of the performance of buildings constructed following these codes allows researchers to examine the efficacy of many aspects of the U.S. model code.

[36] The Haiti quake focused its energy into a few seconds of violent motion close to a highly vulnerable population center. In contrast, the depth of the Chilean quake and its distance from population centers meant that the high-frequency energy had dissipated before it reached the built environment. That said, the shaking in Chile lasted up to two minutes with much of the energy in long-period waves that resonate with tall buildings and critical infrastructure. U.S. researchers will be analyzing how Chile's infrastructure and buildings fared and what can be learned for managing similar risk in the United States, including the Pacific Northwest.

[37] For more information, see CRS Report RL33769, *International Crises and Disasters: U.S. Humanitarian Assistance, Budget Trends, and Issues for Congress*, by Rhoda Margesson.

[38] Section 402 of Title 10, named after former Senator Jeremiah Denton, authorizes shipment of privately donated humanitarian goods on U.S. military aircraft provided there is space and they are certified as appropriate for the disaster by USAID/OFDA. The goods can be bumped from the transport if other U.S. government aid must be transported.

In: Chilean Earthquake of 2010
Editors: J. T. Markou and S. A. Durkin
ISBN: 978-1-62100-184-3


*Chapter 2*

# REPORT ON THE 2010 CHILEAN EARTHQUAKE AND TSUNAMI RESPONSE*

## *United States Department of the Interior, United States Geological Survey*

By the American Red Cross Multidisciplinary Team
Principal investigators: Richard Hinrichs, PhD, CEM,
Lucy Jones, Ph.D., Ellis M. Stanley, Sr., CEM,
and Michael Kleiner, CEM

Open-File Report 2011-1053

**U.S. Department of the Interior**
KEN SALAZAR,
Secretary
**U.S. Geological Survey**
Marcia K. McNutt,
Director
U.S. Geological Survey, Reston, Virginia: 2011

---

* This is an edited, reformatted and augmented version of the United States Department of the Interior, United States Geological Survey publication, Open-File Report 2011-1053.

# 1.0. Executive Summary

In July 2010, in an effort to reduce future catastrophic natural disaster losses for California, the American Red Cross coordinated and sent a delegation of 20 multidisciplinary experts on earthquake response and recovery to Chile. The primary goal was to understand how the Chilean society and relevant organizations responded to the magnitude 8.8 Maule earthquake that struck the region on February 27, 2010, as well as how an application of these lessons could better prepare California communities, response partners and state emergency partners for a comparable situation. Similarities in building codes, socioeconomic conditions, and broad extent of the strong shaking make the Chilean earthquake a very close analog to the impact of future great earthquakes on California. To withstand and recover from natural and human-caused disasters, it is essential for citizens and communities to work together to anticipate threats, limit effects, and rapidly restore functionality after a crisis.

The delegation was hosted by the Chilean Red Cross and received extensive briefings from both national and local Red Cross officials. During nine days in Chile, the delegation also met with officials at the national, regional, and local government levels. Technical briefings were received from the President's Emergency Committee, emergency managers from ONEMI (comparable to FEMA), structural engineers, a seismologist, hospital administrators, firefighters, and the United Nations team in Chile. Cities visited include Santiago, Talca, Constitución, Concepción, Talcahuano, Tumbes, and Cauquenes. The American Red Cross Multidisciplinary Team consisted of subject matter experts, who carried out special investigations in five Teams on the (1) science and engineering findings, (2) medical services, (3) emergency services, (4) volunteer management, and (5) executive and management issues (see appendix A for a full list of participants and their titles and teams). While developing this delegation, it was clear that a multidisciplinary approach was required to properly analyze the emergency response, technical, and social components of this disaster. A diverse and knowledgeable delegation was necessary to analyze the Chilean response in a way that would be beneficial to preparedness in California, as well as improve mitigation efforts around the United States.

By most standards, the Maule earthquake was a catastrophe for Chile. The economic losses totaled $30 billion USD or 17% of the GDP of the country. Twelve million people, or 3/4 of the population of the country, were in areas that felt strong shaking. Yet only 521 fatalities have been confirmed, with 56 people still missing and presumed dead in the tsunami.

The Science and Technology Team evaluated the impacts of the earthquake on built environment with implications for the United States. The fires following the earthquake were minimal in part because of the shutdown of the national electrical grid early in the shaking. Only five engineer-designed buildings were destroyed during the earthquake; however, over 350,000 housing units were destroyed. Chile has a law that holds building owners liable for the first 10 years of a building's existence for any losses resulting from inadequate application of the building code during construction. This law was cited by many our team met with as a prime reason for the strong performance of the built environment. Overall, this earthquake demonstrated that strict building codes and standards could greatly reduce losses in even the largest earthquakes. In the immediate response to the earthquake and tsunami, first responders, emergency personnel, and search and rescue teams handled many challenges. Loss of communications was significant; many lives were lost and effective coordination to support life-sustaining efforts was gravely impacted due to a lack of inter- and intra-agency coordination.

The Health and Medical Services Team sought to understand the medical disaster response strategies and operations of Chilean agencies, including perceived or actual failures in disaster preparation that impacted the medical disaster response; post-disaster health and medical interventions to save lives and limit suffering; and the lessons learned by public health and medical personnel as a result of their experiences. Despite devastating damage to the health care and civic infrastructure, the health care response to the Chilean earthquake appeared highly successful due to several factors. Like other first responders, the medical community had the ability and resourcefulness to respond without centralized control in the early response phase. The health care community maintained patient care under austere conditions, despite many obstacles that could have prevented such care. National and international resources were rapidly mobilized to support the medical response.

The Emergency Services Team sought to collect information on all phases of emergency management (preparedness, mitigation, response, and recovery) and determine what worked well and what could be improved upon. The Chileans reported being surprised that they were not as ready for this event as they thought they were. The use of mass care sheltering was limited, given the scope of the disaster, because of the resiliency of the population. The impacts of the earthquake and the tsunami were quite different, as were the needs of urban and rural dwellers, necessitating different response activities.

**Recommendations for California**

1) Conduct comprehensive exercises (including joint Government, private sector, NGO, emergency responder, and community exercises) before the event.
2) Empower people to be prepared.
3) Educate the population, comprehensively and continually, about what will happen during the event.
4) Emergency and earthquake professionals should work with representatives of print and broadcast media before the disaster to determine how to best serve the community.
5) Emergency plans need to be redundant, flexible, and detailed to handle the unexpected in very large disasters.
6) Recognize the competing personal and professional demands that will be made on staff after a disaster and include this in emergency plans.
7) Organizations need to plan for nonstructural damage and the potential need to evacuate even without structural damage.
8) Recognize vulnerabilities in our communications systems and make comprehensive backup plans to avoid complete communication collapse.
9) Explore mechanisms to encourage building owners to adhere rigorously to existing building codes.
10) Collect all possible data about each disaster when it happens.

The Volunteer Services Team examined the challenges faced in mobilizing a large number of volunteers to assist in the aftermath of a disaster of this scale. One of the greatest challenges expressed was difficulty in communication; the need for redundancy in communication mechanisms was cited. The flexibility and ability to work autonomously by the frontline volunteers was a significant factor in effective response. It was also important for volunteer leadership to know the emergency plans. These plans need to be flexible, include alternative options, and be completed in conjunction with local officials and other volunteers.

The Executive/Red Cross Management Team took a broad look at the impacts of the earthquake and the implications for California. Some of the most important preparation for the disaster came from relationships formed before the event. The communities with strong connections between different government services generally fared well. The initial response and resilience of individuals and communities was another important component.

Communication system failures limited the ability of a central government to assist impacted communities, or to issue tsunami warnings. It also delayed the response since the government did not know (in some case for several days) the impact and needs of local governments. In general, plans for congregate care shelters existed but were little used as most people chose to stay at damaged homes or with relatives. Looting was a surprise to response officials as well as social scientists, but both public and private sector organizations, including NGOs (Non-Governmental Organizations), must consider security for damaged businesses as a priority in California's multihazard planning. Class and ethnic divisions that become heightened during some cases of actual or perceived injustice may also emerge in natural disasters in California.

Several factors contributed overall to the low casualty rate and rapid recovery. A major factor is the strong building code in Chile and its comprehensive enforcement. In particular, Chile has a law that holds building owners accountable for losses in a building they build for 10 years. A second factor was the limited number of fires after the earthquake. In the last few California earthquakes, 60% of the fires were started by electrical problems, so the rarity of fires may have been affected by the shut down of the electricity grid early in the earthquake. Third, in many areas, the local emergency response was very effective. The most effective regions had close coordination between emergency management, fire, and police and were empowered to respond without communication with the capital. The fourth factor was the overall high level of knowledge about earthquakes and tsunamis by much of the population that helped them respond more appropriately after the event.

# 2.0. INTRODUCTION

At some point in the future of the United States, we will be faced with a very large and potentially catastrophic earthquake. The American Red Cross recognized the February 27, 2010, Maule earthquake in Chile as an opportunity to learn and prepare for what we will face in our country after a similar disaster. In partnership with the Multi Hazards Demonstration Project of the U. S. Geological Survey, whose mission is to use hazards science to improve community resilience to natural disasters, the American Red Cross created a delegation of earthquake disaster specialists to go to Chile in order to learn lessons from that earthquake. They brought experts in emergency management and response, volunteer management, medical services, seismologists, and engineers. This report of that delegation on the successes and challenges that Chile faced in responding to their disaster is intended to aid both the American Red Cross and other facets of American society to be prepared for our future disasters.

While the earthquake is inevitable, the magnitude of the human disaster is not. We know that the degree to which we are prepared before a large earthquake will dictate how well we are able to survive and thrive after the event. The challenge to those in disaster prone regions is to understand the consequences of the largest events before they happen, even though, by definition, those events happen so infrequently that they are not in the living memory of most residents. The most effective solution to this problem is to learn from other regions that have experienced similar events. It is essential to appropriately extrapolate what the factors are that correspond between one disaster and another. What are the general lessons we can draw from societies that have experienced a major disaster? What did they do that helped improve their resiliency? In hindsight, what could have feasibly been done to lessen the negative impact of the disaster?

The delegation was structured into five teams: Health and Medical Services, Emergency Management, Volunteer Management, Executive/ Red Cross Management, and Science and Technology (see appendix A). Each team had a different function and role on this trip. The Health and Medical Services team had the goal of understanding the medical disaster response that occurred in Chile after the earthquake. The Emergency Management team sought to get a better handle on the timeline of emergency response, as well as what went well and what went poorly in Chile's response. The Volunteer Management team intended to better understand the role of the Chilean Red Cross and to see how volunteers were handled during the earthquake and tsunami aftermath. The Executive/Red Cross Management Team went to Chile with the intent of broadly observing the disaster there and noting similarities to events that the American Red Cross, its State response partners, and other organizations in California might face. The Science and Technology team served as a resource to the other teams to help understand the science and engineering sides of the earthquake and tsunami. All teams went to Chile in order to learn from the strengths and weaknesses of the Chilean response, with the goal of taking lessons back to the United States to increase our preparedness.

The following sections detail the findings of the different subject areas addressed by the delegation.

# 3.0. Science and Technology

The Science and Technology Team compiled the scientific and technical findings about the Chilean earthquake to provide a context for comparing this event to future events in California and elsewhere. The Team met with representatives of Chile's National Seismological Service and university-based structural engineers and visited multiple sites of damage.

## 3.1. The Earthquake

The moment magnitude 8.8 Maule earthquake began at 3:34 a.m. on Saturday, February 27, 2010. The epicenter of the earthquake was offshore from the Maule Region, approximately 105 km (65 mi) north-northeast of Chile's second largest city, Concepción (U.S. Geological Survey Earthquake Program, 2010a). Since the earthquake was in the middle of the night on a weekend, most people were asleep in their homes. Injury due to falling debris or damaged roadways, bridges, and highways was minimized due to the time and day of the earthquake. 521 people were killed, and 56 are still considered missing. A brief summary of Chile's losses in this event can be seen in table 1.

**Table 1. Summary of damages and losses from February 27, 2010, earthquake and tsunami (Comité de Emergencia, written communication, 2010)**

| | |
|---|---|
| Deaths | 521 people |
| Missing Persons | 56 people |
| Homes destroyed or damaged | ~370,000 (11% of the total in the area) |
| Hospitals destroyed or damaged | 73 |
| Schools destroyed or damaged | 3,049 schools, housing 1.25 million students |
| Bridges destroyed or damaged | 221 |
| Villages and rural and coastal communities affected | More than 900 |
| Estimated economic losses | U.S. $ 30 billion. 17% of GDP |

A major tsunami struck the coast nearest to the epicenter within minutes of the *M*8.8 earthquake. The earthquake occurred in the Chilean subduction zone, where a section of the earth's crust called the Nazca Plate is subducted under the South American Plate.

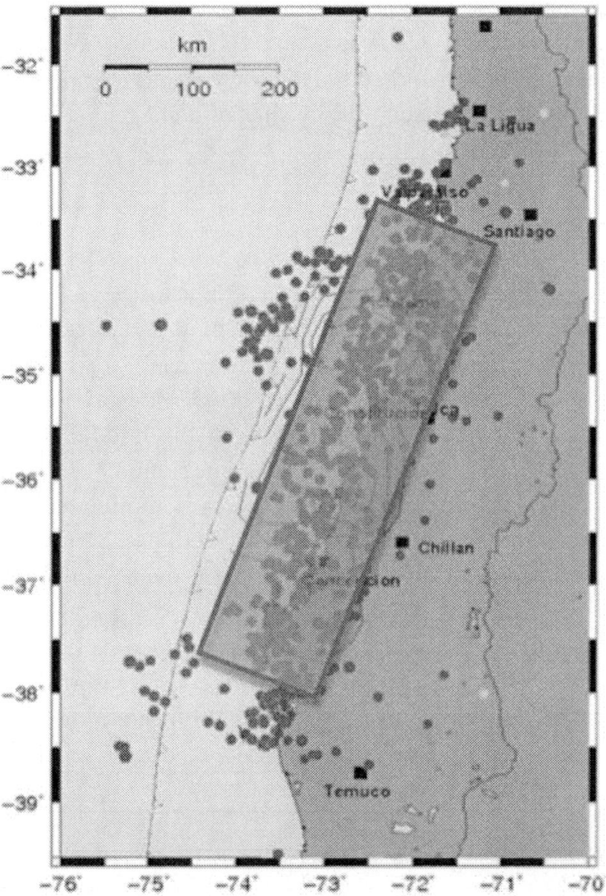


Figure 1. A map of *M*>4.7 aftershocks from the February 27, 2010, Chile earthquake. The epicenter of the *M*8.8 earthquake is marked with a red star. The contour lines represent the amount of displacement at the boundary between the Nazca and South American Plates. Maximum displacement reached values of 12---14 m (National Center for Earthquake Information). The red box outlines the fault plane that ruptured in the earthquake. (Image courtesy of Sergio Barrientos).

The size of an earthquake is proportional to the area of the fault plane that moves in an earthquake, multiplied by the amount one side of the fault moved with respect to the other. The section of fault that moved in this earthquake (see fig. 1) was 450-500 km long (280-310 miles) and around 100 km wide

(60 miles). The rupture began at the epicenter and propagated both north, towards the capital city of Santiago, and south of the second largest city of Concepción.

**Table 2. A description of Modified Mercalli Intensity categories. These categories are based on observation, and not measurements made by instruments (U.S.G.S. Earthquake Program, 2010b)**

| Intensity | Description of Shaking/Damage |
|---|---|
| I | Not felt except by a very few under especially favorable conditions. |
| II | Felt only by a few persons at rest, especially on upper floors of buildings. |
| III | Felt quite noticeably by persons indoors, especially on upper floors of buildings. Many people do not recognize it as an earthquake. Standing motor cars may rock slightly. Vibrations similar to the passing of a truck. Duration estimated. |
| IV | Felt indoors by many, outdoors by few during the day. At night, some awakened. Dishes, windows, doors disturbed; walls make cracking sound. Sensation like heavy truck striking building. Standing motor cars rocked noticeably. |
| V | Felt by nearly everyone; many awakened. Some dishes, windows broken. Unstable objects overturned. Pendulum clocks may stop. |
| VI | Felt by all, many frightened. Some heavy furniture moved; a few instances of fallen plaster. Damage slight. |
| VII | Damage negligible in buildings of good design and construction; slight to moderate in well---built ordinary structures; considerable damage in poorly built or badly designed structures; some chimneys broken. |
| VIII | Damage slight in specially designed structures; considerable damage in ordinary substantial buildings with partial collapse. Damage great in poorly built structures. Fall of chimneys, factory stacks, columns, monuments, walls. Heavy furniture overturned. |
| IX | Damage considerable in specially designed structures; well---designed frame structures thrown out of plumb. Damage great in substantial buildings, with partial collapse. Buildings shifted off foundations. |
| X | Some well---built wooden structures destroyed; most masonry and frame structures destroyed with foundations. Rails bent. |
| XI | Few, if any (masonry) structures remain standing. Bridges destroyed. Rails bent greatly. |
| XII | Damage total. Lines of sight and level are distorted. Objects thrown into the air. |

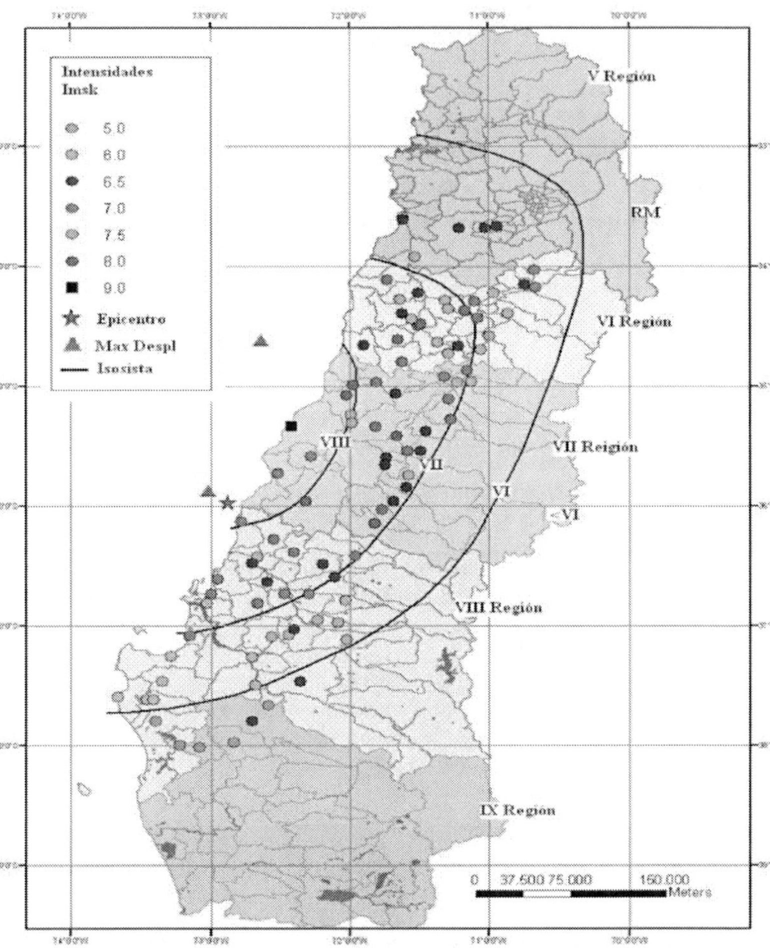


Figure 2. Isoseismal map (intensity map) produced after inspecting 98 localities throughout the length of Chile. According to the 2002 Chilean Census, 12,202,858 people live in this broad area, representing 79.5% of the population of the country. The locations visited by Astroza and coworkers. are distributed over the length and breadth of the country achieving an almost complete coverage of the earthquake damage zone. Villages at the foot of the Andes were not visited; many of these areas are far from the rupture plane and no damage was reported. Some villages along the coast that were affected by the tsunami were also not visited. It was difficult to differentiate what damage was caused only by ground motion, which is a fundamental piece of information when determining the degree of seismic intensity. (Image from Astroza and others., 2010).

In spite of the large magnitude, the intensity of shaking was lower than in some other similarly sized earthquakes because the fault plane was mostly offshore. The Modified Mercalli Intensity scale (MMI) describes the intensity of shaking in an earthquake on a scale of I to XII, the categories of which are defined in table 2. A map of the intensities experienced in the Chile earthquake is shown in figure 2. The highest intensity, MMI IX, was recorded at the town of Constitución on the coast, just north of the epicenter. Concepción received Intensity VIII shaking, while intensity VI was recorded for the capital of Santiago. By comparison, the most intense shaking in the 1994 Northridge, California, earthquake was MMI IX in much of the western San Fernando Valley, while Pasadena received MMI VI shaking in that event.

The Modified Mercalli Intensity scale, composed of different levels of intensity that range from barely felt shaking to catastrophic destruction, is split into 12 increasing levels of damage, which are ranked by Roman numerals. It does not have a mathematical basis; instead it is an arbitrary ranking based on observed effects of ground shaking.

The Modified Mercalli Intensity value assigned to a specific site after an earthquake provides a more meaningful measure of severity to the nonscientist than the magnitude of an earthquake. Intensity refers to the effects actually experienced at that location, rather than earthquake magnitude, which does not describe the damage that was experienced.

## 3.2. The Tsunami

A tsunami is a large series of waves that travel through the ocean, which can cause great destruction upon impact at the shoreline. Tsunamis travel across the ocean at speeds comparable to commercial airplanes (500-650 miles per hour), but waves slow down when they approach the shoreline. Subduction zone earthquakes produce the largest tsunamis.

The change in sea height recorded on the coast of Chile on February 27, 2010, is shown in fig. 3. Note how the tsunami waves continue for more than 12 hours. The later pulses decay with time, but the tsunami signal is superimposed on the tides. For this earthquake, high tide occurred about 3-8 hours after the earthquake (depending on the location of the tide gauge), so the highest water occurred in many locations around 7 a.m-1 p.m.

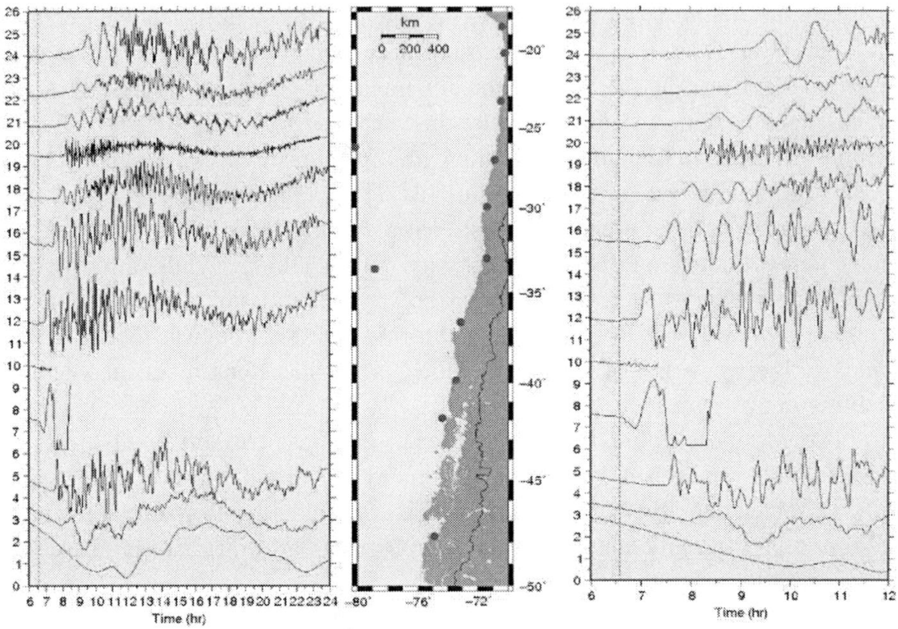

Figure 3. Water level recordings for tide gauges along Chilean coastline. The red dots on the map in the center show the locations of the hydrographs. The graphs on the left show water levels over 18 hours after the earthquake (the red line is the time of the earthquake). The graph on the right is on an expanded time scale to see the first 6 hours. Note that sea level fluctuations occurred all day, not just immediately following the earthquake. The tsunami hit as neap tide was approaching, which is when the lowest---low tides occur. Therefore, the largest tides came in when high tide was approaching, 3---8 hours later in some places. (Image courtesy of Sergio Barrientos).

## 3.3. Impact of the Earthquake

Estimates from the Chilean Government suggest that over 370,000 housing units were damaged or destroyed in the February 27th earthquake. Table 3 lists the Government of Chile's estimates of housing damage (as of March 29, 2010) (GEER, 2010). In 6 of 15 of Chile's regions, a "state of catastrophe" was declared. The Government of Chile states that more than 12.8 million people in the most affected regions were impacted by the earthquake and resulting tsunami.

An earthquake can damage structure (for example, buildings), infrastructure (for example, lifelines and roadways), and non-structural entities (for example, furniture inside of a building). Chile has very strong building codes. Structures in areas that experienced strong shaking had less damage than they would have if building codes were worse. Buildings in Chile are built to be rigid, as opposed to the more flexible construction used in southern California structures. The apartment building in figure 4 (the Alto Rio building in Concepción) did not collapse in a pancake-style, but rather fell over during the shaking. This structural response is a product of rigid building design.

**Table 3. Estimate of housing damage (as of March 29, 2010). From the Geo---Engineering Extreme Events Reconnaissance Team Report (GEER, 2010)**

| | Housing Units Destroyed | Housing Units Significantly Damaged | Housing Units with Minor Damage | Total Housing Units |
|---|---|---|---|---|
| Coast | 7,931 | 8,607 | 15,384 | 31,922 |
| Urban Adobe | 26,038 | 28,153 | 14,869 | 69,060 |
| Rural Adobe | 24,538 | 19,783 | 22,052 | 66,373 |
| Multi-unit Public Housing | 5,489 | 15,015 | 50,955 | 71,459 |
| Multi-unit Private Housing | 17,449 | 37,356 | 76,433 | 131,238 |
| Total | 81,444 | 108,914 | 179,693 | 370,051 |

(Source: Government of Chile)

Chile's strong building codes saved many lives in this earthquake and are probably the most significant factor in the low mortality level. Simply because a country has strong building codes, however, does not mean that they will be enforced. Another possible factor in Chilean buildings' widely successful response to the earthquake is Chilean Law 458: the General Urbanism and Construction Law. This law requires compensation to be given by the original construction company to those who suffer loss in the wake of partial or complete destruction of a building. Compensation will only be given out if building codes were not adhered to. The General Urbanism and Construction Law discusses the liability of the construction company. If this company followed and respected the building codes, they are not the responsible for providing compensation. If they did not build according to the building code, they are the responsible for the damages in the building that were due to a failure during the construction process.

Article 18 of the law declares that the original owner is responsible for all severe damage that occurs within the first 10 years of the building's existence. (The law scales, so the owner is not liable for certain damages after a given period of time.) The law does not discriminate whether these failures are due to errors in the construction process or from adjustments made after the building was constructed. If the building (apartment or house) is insured, the insurance pays for the people who were injured. These compensations may last up to 30 years.

Figure 4. The collapsed Alto Rio apartment building in Concepción. Eight people were killed when this building fell over.

In the region that experienced the highest intensity shaking, infrastructure, including roads, bridges, airports, ports, utilities, and communication networks, sustained significant damage. The most severe damage occurred along the coast and in parts of Chile's central valley. The earthquake disrupted the major industries of the region (fishing, shipping, mining, refineries, forestry, winemaking, and agriculture). The Chilean Government estimates economic losses to be approximately $30 billion USD, which is 17% of Chile's GDP (Oral communication, Comité de Emergencia, 2010). Restoration of telecommunication and electricity systems varied greatly. Shaking induced electric shutoffs in some power stations; this had a domino effect and shut down the entire electric system within about 30 seconds of the start of the earthquake. We received reports of power being restored to many areas within 2-3 days. Telecommunications were also rapidly restored in some areas. In

Concepción, we received reports of cell phone use being restored within two hours. Based on reports from many individuals we spoke with, water, sewage, and transportation damages took much longer to repair. Prolonged damages can significantly raise business interruption losses and strongly impact overall recovery time.

It appears that there were fewer fires following this earthquake relative to the population exposed to strong shaking than has been seen in California. It is not possible to be quantitative about this, because none of the people we spoke to knew of any documentation of the number or extent of fires. It is clear that there were fires, some significant, such as the large fire in a chemical factory outside of Santiago. It is also clear that there were no conflagrations that could not be controlled. Possible contributing factors include the complete shutdown of the national electricity grid within the first 30 seconds of shaking and the widespread use of propane for cooking so that gas pipelines were limited outside the cities.

## 3.4. Comparing the Chilean Earthquake to the January 12, 2010, Haiti Earthquake and to Large Earthquakes on the San Andreas Fault in California

Figure 5 maps the distribution of intensity for three earthquakes, the February 27, 2010, *M*8.8 Chilean earthquake, the January 10, 2010, *M*7.0 Haiti Earthquake, and a hypothetical southern California earthquake, the *M*7.8 ShakeOut scenario. All three are shown at approximately the same scale. The ShakeOut and Haiti earthquakes are both on vertical, strike-slip faults that reach the surface of the earth in populated areas. The Chilean earthquake occurred on a fault that was almost horizontal, that comes to the surface in the seafloor. The area of the fault determines the magnitude of the earthquake. The area of the Chilean fault is 20 times the area of the ShakeOut fault and almost 200 times the area of the fault in Haiti. Since the Chilean earthquake began in the middle of the fault and ruptured in both directions, the duration of the rupture was shorter than if it had begun at one end. Estimates of the duration are about two minutes, which is comparable to the ShakeOut earthquake (which ruptures from the southern end of the San Andreas northward). The Haiti earthquake rupture lasted for about 15 seconds.

Thus the Haiti earthquake had a lower level of shaking for a shorter period of time than either Chile or the ShakeOut. The extremely high death toll and level of damage in that earthquake is directly attributable to the poor quality of buildings in the area. The Haiti experience, therefore, does not provide a basis for comparison to either the Chilean or ShakeOut earthquakes.

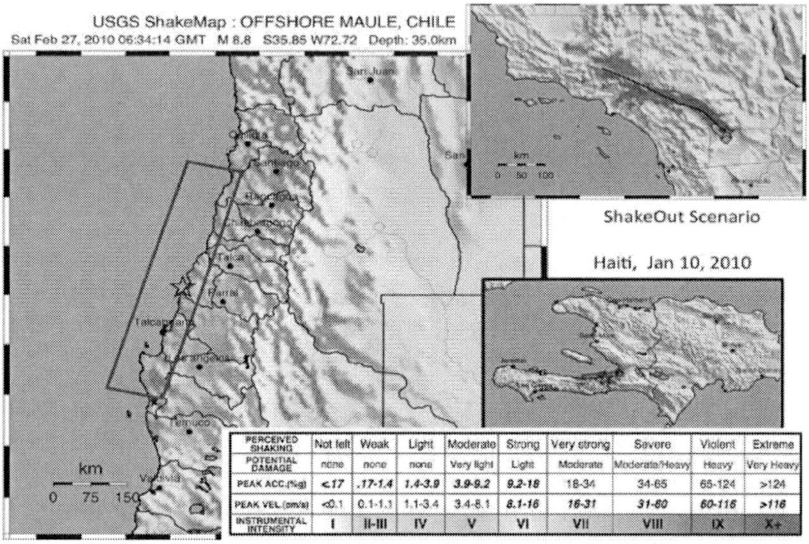


Figure 5. Maps comparing the intensity of shaking during three earthquakes: the February 27, 2010, *M*8.8 Chilean earthquake (left), the January 10, 2010, *M*7.0 Haiti earthquake (lower right), and a hypothetical southern California earthquake, the *M*7.8 ShakeOut scenario (upper right). Colors show the level of shaking as Modified Mercalli Intensity (Images courtesy of USGS).

Compared to the expected San Andreas earthquake in California, the Chilean earthquake occurred on a longer and wider fault, with fewer people near the fault. This means the type of damage caused by being located near the fault (from very high frequency shaking) would be more prevalent in California compared to the Chilean earthquake. The type of damage that comes from the very large, low frequency waves of a great earthquake will dominate the Chilean damage. This is the type of damage that will be more prevalent in a big San Andreas earthquake than in other previous California events, such as the Northridge and Loma Prieta earthquakes.

These examples highlight the fact that magnitude alone does not determine what the experience of an earthquake will be. The intensity of shaking at any one moment does not grow significantly at the largest magnitudes. Rather, the duration of the shaking increases, and the area exposed to the strong shaking gets larger. The population exposed to the shaking, however, is equally important. Many of the Chilean officials interviewed by this delegation said that they were surprised by the extent of the devastation because they know they had already experienced the largest earthquake in recorded history (the magnitude 9.5 event in 1960), and they thought that they should have been able to handle anything else that came along. However, the 1960 event was in the southernmost part of the country, which has a limited population, and did not approach the population exposure of the 2010 earthquake, which was located near Chile's second largest city. Similarly, California should recognize that a magnitude 7.4 on a fault that extends under most of Los Angeles would be much more devastating than the same size earthquake on a fault located in a more rural area. Magnitude, proximity of population, and local soil conditions all play a role in determining the level of damage in an earthquake.

### 3.5. Science and Technology Key Findings

1) The law that requires building owners to compensate for damages caused by failure to adhere to the building code was repeatedly cited as an incentive for more conservative construction. Some method to encourage personal responsibility by building owners in California would increase our resiliency.
2) Lives were saved because most of the local population knew about the risk of tsunamis and moved quickly away from the coast after the earthquake. It is not clear if lives were lost because not everyone realized that a tsunami is not one wave but a series superimposed on the tides thus the highest surge could be several hours later when the tide is highest. Comprehensive and continuing education about earthquake and tsunami risks could save lives in California.
3) 3. Not many fires were triggered by this earthquake. The degree to which this was affected by electrical systems that were shut down before strong shaking arrived needs to be better understood.

## 4.0. Emergency Management

The Emergency Management Team evaluated earthquake and tsunami preparedness and response by collecting information through meetings with National and local government officials, Chilean Red Cross personnel, the United Nations International Coordination Officer, and *Bomberos* (firefighters), and through individual interviews with disaster survivors. These meetings took place in the cities of Santiago, Talca, Constitución, Concepción, Talcahuano, Tumbes, and Cauquenes between July 19 and July 27, 2010. Pre-scheduled meetings with government and Chilean Red Cross personnel included presentations and group discussions, each with question and answer periods. During meetings, interviews, and conversations the Emergency Management Team sought to collect information on all phases of emergency management actions (preparedness, mitigation, response, and recovery), and determine what did and did not work.

## 4.1. Government and Structure of Emergency Management in Chile

### *4.1.1. Chilean Political Structure*

Chile has an area of approximately 303,000 square miles, twice the size of California, but has a population of only 16.8 million, 6.5 million of whom live in Chile's capital of Santiago. The earthquake's greatest impact was on the most heavily populated central region of the country, which includes the second largest urban area of Concepción-Talcahuano (population 840,000) as well as Santiago. Chile is one of the more prosperous nations in South America, with an annual GDP of $173 billion and an annual growth rate of 3.2% based on mining, manufacturing, and exports of copper, seafood, and agricultural produce and wood products. Per capita income is $14,688, and the adult literacy rate is 96%.

Chile is a republic with executive (president), legislative (bicameral legislature), and judicial (Constitutional Tribunal, Supreme Court, courts of appeal and military) branches. Administratively, Chile is divided into 16 regions under the authority of *intendentes* appointed by the president. Within each region are provinces, each with an appointed governor; provinces are divided into districts or municipalities administered by popularly elected mayors. Within this governmental structure, the principal emergency response organizations are Officina National de Emergencias, Ministerio de Interior (ONEM), firefighters, and armed services.

### *4.1.2. Officina National de Emergencias, Ministerio de Interior, or National Office of Emergencies, Ministry of the Interior (ONEMI)*

ONEMI is the agency responsible for "Civil Protection, understood as protecting people, property and environment in a situation of collective risk, whether naturally occurring or generated by human activity is performed in Chile by a system composed of Organizations, Services & Institutions both of the public and private sector, including voluntary agencies and community organizations under the coordination of the National Emergency Office of the Ministry of Interior, ONEMI" (March 12, 2002).

ONEMI (see fig. 6) is roughly comparable to the Federal Emergency Management Agency (FEMA) in the United States and has some common functions such as training, promoting preparedness and hazard mitigation, and coordinating with other agencies of government and the private sector in emergency response and recovery. But ONEMI also has responsibility for alert and warning, functions performed in the United States by the U.S. Geological Survey for earthquakes and the National Oceanic and Atmospheric Administration's (NOAA) National Weather Service for Tsunamis.

In addition, as Chile's main civil protection agency, ONEMI performs some of the coordination and mutual aid functions associated with state government in California through emergency response committees at the regional, provincial, and municipal levels.

Figure 6. American Red Cross Delegates meet with ONEMI officials in Santiago.

### *4.1.3. National Volunteer Fire Services*

Fire services in Chile are provided by 38,000 volunteer firefighters (*Bomberos*), a fact that astounded many of the members of the U.S. emergency management team who traditionally associate volunteer fire fighting with small towns and rural areas. There are no paid professional firefighters in all of Chile, including Santiago, an international city of 6.5 million inhabitants. There is support from the National government, which equips local fire departments in a total of 345 districts and operates a national fire academy to train Chilean firefighters as well as those of other South American nations. The only paid functionaries associated with fire services are dispatchers and drivers. The President of the *Bomberos* has administrative responsibility for regional fire authorities. The regional fire authorities serve as coordinators of fire services and act as mutual aid providers at the provincial and municipal levels.

The responsibilities of Chilean fire agencies are similar to those handled by firefighters in the United States. They are responsible for fire suppression, hazardous materials management, emergency and medical response, and search and rescue (see fig. 7). In addition, they conduct safety inspections of buildings following major emergencies, a function performed by volunteer engineers and building officials in California under a program (ATC-20 Safety Assessment Program) administered by the California Emergency Management Agency.

In spite of their unpaid status, fire agencies in Chile appear to have little difficulty in recruitment and retention of personnel. The long tradition of volunteer firefighting in Chile spanning 160 years has witnessed the evolution of mechanisms to ensure adequate numbers of new volunteers. These mechanisms include a youth program that begins firefighting training at the age of 12 and results, at age 18, in sufficient prerequisite training to be admitted to the fire academy. Further, there is significant cultural prestige associated with being a firefighter such that recruits are drawn from all social classes, including professionals. A testament to the status of firefighting in Chile is the fact that there is not a law mandating the release of *Bomberos* from their paid jobs, only the sense of community responsibility on the part of employers.

Figure 7. Chilean firefighters distribute water to a local community after the earthquake.

#### *4.1.4. Armed Forces*

Chile's armed forces are subject to civilian control under a Ministry of Defense and the President of the Republic. Commanders of the armed forces are appointed by the President for 4-year terms and serve at the discretion of the President. The Chilean Army consists of 45,000 soldiers and is organized into seven divisions located at bases throughout the country and headquartered in Santiago. Chilean armed forces also include an Air Brigade in Rancagua and a Navy. A key responsibility of the Navy's Hydrographic and Oceanographic Service is to monitor the Pacific Tsunami Warning Center and issue tsunami warnings for Chile.

## 4.2. Emergency Management Findings

The following paragraphs summarize the key points pertinent to emergency management. During meetings and interviews, the emergency management team sought to collect information on all phases of emergency management (preparedness, mitigation, response, and recovery) and determine what worked well and what did not work. Below is a summary of the data received related to each response phase.

#### *4.2.1. Preparedness / Mitigation*

Several of those who discussed preparedness with the Emergency Management Team expressed an opinion that Chile was not adequately prepared for the effects of a disaster the size of which occurred on February 27, 2010. This opinion was shared by National and local government personnel, the Chilean Red Cross, as well as individual disaster survivors that were interviewed.

ONEMI acknowledged that the communications systems and plans that existed at the time of the event clearly failed, meaning they did not perform as expected, leaving the agency initially unable to communicate with the impacted areas. ONEMI indicated that the absence of any major disasters in Chile for over 20 years and the resulting reduction in their profile and funding were contributing factors to this failure. An additional element was the belief by many that Chile had already experienced and survived the most powerful earthquake of the 20th Century, a *M*9.5 in August of 1960, and therefore was at little risk for another major earthquake.

Fire department officials in the town of Concepción also described challenges that impacted the execution of their and other response efforts. Two significant issues in plan coordination were noted with the fire response. In Concepción, even though there was a plan for how off-duty *Bomberos* would first ensure their families' safety and then respond to the earthquake, on-duty fire crews were not part of these plans. In the aftermath of the quake, this led some of the on-duty personnel to deviate from their assigned response duties in order to check on their families and property. Second, their plan identified a specific location outside of the tsunami threatened area as a muster point for fire crews and equipment. Due to damage to roads, responders were unable to reach these muster point. Since no back-up site was identified in the plan, many *Bomberos* were unclear on their next steps.

Preparedness at the local level appears to have been effective. For example, the ONEMI Regional Emergency Management in Talca reported that most coastal communities practiced evacuating the coastal area during drills every other month. This activity contributed to the high level of successful evacuation achieved by the communities ultimately hit by the tsunami.

A number of mitigating factors appear to have limited the human losses due to the earthquake and tsunami. During several meetings the Emergency Management Team learned about the "earthquake culture," or "culture of preparedness" that has developed in Chile based on the history of regular, and at times very large, earthquakes. This "culture" proved to be most valuable in the tsunami-hit areas; the generation that survived the 1960 *M*9.5 earthquake had instilled a belief that anytime an earthquake was large enough to make it difficult to walk, people should quickly move to higher ground. During meetings with government officials, Chilean Red Cross, and individuals who were in the area when the tsunami hit, all expressed a belief that most individuals who were physically able to seek higher ground, did so. In the coastal areas, the population knew through previous exercises and information shared about past earthquakes that they needed to get to higher ground. A common assertion was that casualties on the coastal areas were low because the population acted independently and did in fact seek higher ground. Another cultural mitigating factor is the community spirit of the Chilean people. The responders and individuals interviewed in the impacted areas described how people shared resources such as food and shelter with family and neighbors. The fire chief in the town of Constitución described how people shared food, water, and shelter in the days before outside government resources arrived. The first outside food to arrive in Constitución came two days after the event and was brought in by family members who drove from the city of Talca. Although three disaster shelters were opened in the first few days, they quickly closed as the displaced population moved out, as they had primarily sought housing with family or friends.

Chile's method for delivering food to their needy populations also helped to mitigate potential food shortages after the earthquake. Food is stored in "4x4" boxes in a central distribution warehouse. These boxes contain non-perishable items that can feed four people for up to four days. During non-disaster times these boxes are used to feed people who cannot afford to buy their own food. During the disaster, local communities, supported by ONEMI at the National level, were able to access and expand this existing system to deliver food to the earthquake and tsunami impacted areas.

#### *4.2.2. Response*

The initial response to the earthquake and tsunami varied based on location and several other factors. We will discuss local and community response, including the Chilean Red Cross, followed by the National/Regional response.

#### 4.2.2.1. Local/Community Response

In the first hour and days following the earthquake and tsunami(s) the local and community response was dependant on local resources, local first responders, and individuals within the community—with very little outside assistance. Descriptions of the length of time it took to receive outside assistance varied depending on the location and the individual you were speaking with. Perspectives indicated that the local government response to the quake focused initially on search and rescue, medical support, and provision of food and water. However, for example, fire officials in Concepción said that many fire stations were damaged, which significantly impacted the response (see fig. 8). Effectiveness of the limited fire department resources was further complicated in Concepción by looters who in some areas barricaded the streets, making it difficult to put out fires. Fire officials also reported that some fire and ambulance personnel left their shifts to check on their family and property.

Figure 8. Damaged Firehouse in Talcahuano, near Concepción.

Members of the Emergency Management Team received repeated reports of the independent actions by residents within the impacted areas in response to the quake and tsunami. Individuals assisted others by rescuing them from damaged buildings, providing first aid, and providing and sharing food, shelter, and other resources. Since there was, at least initially, little communication with ONEMI or other agencies outside the heavily impacted areas, this individual and community response was vital and critical.

#### 4.2.2.2. The Chilean Red Cross Response

Prior to this earthquake and tsunami, the Chilean Red Cross' role had primarily been medical emergency response (paramedics); they had not been considered a "major player" in disaster preparedness and response. Chilean Red Cross personnel reported that their disaster preparedness role has been increasing since 2006. They expressed the belief, shared by many others with whom we met, that the damaging effects of this event were larger than anyone had anticipated. When the earthquake occurred, National Chilean Red Cross in Santiago was invited to participate at senior level meetings with the National government within three hours. After three days, Chilean Red Cross withdrew their participation because they felt they were not getting useful information and that this collaboration limited their ability to respond effectively using the resources that were becoming available to them. The Chilean Red Cross decided to focus instead on local and regional activities. Unlike the American Red Cross (ARC), the Chilean Red Cross is not responsible for congregate care sheltering. Local government has that responsibility and is supported by non-governmental organizations (NGO's) as necessary.

The Chilean Red Cross focus on response at the local level initially occurred without any coordination with the national government. Chilean Red Cross reported that very little information came from the government in the first few days. Their immediate priorities were focused on assessing impact, helping people, and providing temporary housing. They distributed coats, tents, food, and water and helped with basic recovery services.

Chilean Red Cross reported that very few shelters were opened. Most people remained with their property or stayed with friends or family. Chilean Red Cross sent volunteer assessment teams into the communities to first assess needs, and then later returned to deliver basic food and other items. Some of these deliveries were made house by house, some through bulk distribution sites. The Chilean Red Cross delivered 5,071 food boxes that contained a supply for 5 people for 15 days. They also assisted in the distribution of other government-supplied food boxes. Chilean Red Cross also assisted with providing health and sanitation support services.

Figure 9. A truck sells food to displaced persons in *Mediasaguas* 'village' outside of Constitución.

During a visit to a temporary housing village outside of Constitución where 170 families were housed, residents informed our team that their first assistance after the tsunami came from the Chilean Red Cross. Also while at the village, the Emergency Management Team observed a food vendor drive through the village with fruit, vegetables and fish for sale to residents (see fig. 9). Local Chilean Red Cross officials told us that these deliveries were a daily occurrence and alleviated the need to drive, or find a ride to travel the 2.5 km to town to purchase food.

The International Federation of Red Cross and Crescent Societies (IFRC), the German Red Cross, the Spanish Red Cross, and others supplemented the Chilean Red Cross response. Overall, the quality of the Chilean Red Cross response to the earthquake, which all observers agreed was crucial, positively influenced the reputation of the organization in the eyes of the Chilean government and amongst the public.

#### 4.2.2.3. National Government Response

The initial National response to the event was marked by confusion and a lack of clear situational awareness. Immediately following the earthquake, there was confusion about whether or not a tsunami would follow. It was reported that the outgoing president, based on misinformation from the Navy's Hydrographic and Oceanographic Institute, initially announced that there was not a threat of tsunami, and then minutes later warned that there was a threat. Few people in the tsunami-impacted areas heard either announcement since communications with Santiago were lost. The confusion over a tsunami warning (or lack thereof) was not a contributing factor in how coastal communities fared. According to officials from the City of Talca, the first wave hit before a tsunami warning could have been issued.

Several individuals interviewed reported that the outgoing president was also slow to involve the military in the response. Although the military is not normally involved in civilian emergencies in Chile, the president can declare a "catastrophic event" and utilize the military for disaster response. Responders expressed the opinion that the decision to declare a catastrophic event in this case took longer than would have been preferred. Many officials believed that looting became more widespread due to a delay in mobilizing the army to assist with local law enforcement and security. When the military began to arrive more than 48 hours after the event, the looting quickly subsided.

Reports and interviews, particularly discussions with United Nations officials, suggested that the government could have taken advantage of offers of assistance from other nations more quickly. This was a source of frustration for many international aid and coordination organizations. Many nations and international groups offered to send assistance or to play their traditional role of coordinating the international response; the repeated reply from the Government of Chile was that they were not needed at that time. When the government eventually asked for international assistance, several organizations reported that they were surprised by what was requested. The initial list was very limited and was not modified until the new government assumed power on March 11.

The perception of local officials was that the National response improved after the initial days of the event, and especially after the installation of the new government. The new president formed an emergency committee to oversee the response and recovery at the national level (see fig. 10). The committee brought together government officials with the goal of overseeing the response and recovery. In the committee's plan, the military played a key role in this response strategy.

The military role was divided into two stages. The goal of the first stage was to control looting and assist local police in the heavily impacted areas. The goal of the second stage (which began on March 31, 2010) was to assist with logistical support, debris removal, and construction of temporary housing. In order to meet these goals, the army was given administrative control of the National response.

Figure 10. American Red Cross team meeting with Cristobal Lira Ibañez, Executive Secretary of the Comité de Emergencia (President's Emergency Committee).

The Army deployed and operated in cooperation with other ministries and organizations. They report to the Commander-in-Chief of the Army, who is located in Santiago. The Army also provided communications for the emergency response effort until public utilities could be restored. According to the fire chief of Constitución, the army arrived on day five and their presence had an immediate calming effect on the city. The army established order and took control of operations. This included taking control of the remaining shelters, erecting field hospitals, and providing food and water to survivors.

### *4.2.3. Shelter / Feeding / Hydration*

Of particular note is that the use of congregate care shelters was very limited in this event. This is in spite of the fact that estimates of upward of 800,000 people were displaced by the earthquake and tsunami. Several communities opened shelters that were not used at all or were used on limited bases for only a few days. People either sheltered in or near what remained of their dwelling, went to stay with family members, friends, or found shelter in other areas within or outside of the impacted areas.

The government utilized several methods to assist those displaced by the earthquake. For individuals and families who required shelter, the government built what they call *Mediasaguas* (see fig. 11). These are 3x8 meter wooden structures (cabins). It should be noted that the *Mediasaguas* are traditionally used in Chile to house portions of their homeless population, making them initially less than acceptable for those displaced by the earthquake and tsunami.

In the areas where the army established order and took control of operations, it included taking control of the remaining shelters and providing food and water to survivors.

The government provided food boxes to the population. The existing food box system normally used for the needy was augmented and used for disaster survivors. Delivery was made via all available routes, including helicopters. Some food boxes were delivered to bulk distribution sites; others were delivered directly to homes in remote locations by the military and other non-governmental organizations such as the Chilean Red Cross.

Caches of food boxes were warehoused by ONEMI at the national, regional, and local levels. Each municipality had a small store of food boxes at the city hall. These boxes were rotated prior to the earthquake through periodic distributions to the poor.

Figure 11. *Mediasaguas* 'village' outside of Tumbes.

### *4.2.4. Recovery*

The new president charged the Emergency Committee with three tasks:

1) Provide basic needs such as food, power, and sanitation within one month.
2) Ensure that everyone had a roof over his or her head before winter (June 1 was the target date)—planning for this began immediately.
3) Find definitive solutions to deal with housing, debris removal, demolition, and restoration of basic services.

It was reported that each of these tasks was progressing as designed. Most utilities were restored well before the one-month goal. In the City of Talca, some water service was restored within a week, although as of July 2010, up to 10 percent of the city still had no water service.

Since the earthquake and tsunami, the government has built approximately 65,000 *Mediasaguas* (as of July 2010). The wooden components of these structures were manufactured at several locations, transported to the impacted area, and then assembled by the army, volunteer agencies, or the survivors themselves. The majority of these structures were placed on land next to a damaged or destroyed home. Others were built in large camps, called *Aldeas* (villages). Most of the inhabitants of the *Aldeas* were those who lost everything due to the tsunami. Communal toilets and showers were installed as the camps were constructed. Electricity was provided to the inhabitants much later. The survivors living in the *Aldea* outside of Constitución reported that they received electricity in May. These buildings and camps are meant to be temporary.

Land for these *Aldeas* was secured from public and private sources. The *Aldea* near Constitución was constructed on land owned by an adjacent sawmill. The Chilean Navy provided land for the *Aldea* near Tumbes, a small town on the peninsula north of Talcahuano.

For longer term housing, other programs have been developed. Those who had repairable damage to their home were eligible for an $800 grant to buy tools and supplies to repair their own homes. About 50,000 homes were repaired through this program. Others who need to rebuild completely are eligible for $25,000-$30,000 in the form of vouchers that can be utilized to rebuild a house following certain government requirements. The government plan is to organize recipients of the vouchers into the form of a co-op so that they can secure better terms from the local building contractors.

The National central government, through the Ministry of Housing, is in the process of developing plans to redesign communities in the tsunami-impacted areas. The government has worked with architects, planners, and engineers to develop plans that will permanently remove housing from tsunami-threatened areas, and in their place will create parks and other non-residential uses for this land. Several prototypes have been developed and local communities are being asked to choose the preferred plan for their community.

The Government of Chile is attempting to improve emergency response capabilities following this disaster. ONEMI personnel reported that since the quake, their intra-agency structures and processes are being updated. They acknowledge that their plans and systems in place at the time of the quake were not sufficient to meet the challenges of this event. According to the Director of ONEMI, improvements will include new legislation, possible elevation of ONEMI to a higher level in the government, and strengthened international planning, (including participation in California's 2010 statewide ShakeOut earthquake exercise).

The Chilean Red Cross also told our team that they are utilizing lessons learned during the event to improve their capabilities. They are continuing to increase their role in disaster preparedness and response. They also feel that this event has led to an opportunity for a stronger relationship between Chilean Red Cross and ONEMI. The new government has included Chilean Red Cross in strategy discussions, and they now feel that ONEMI is listening to them. The new president has also proposed legislation giving ONEMI more responsibilities that, for the first time, include the role of Chilean Red Cross.

## 4.3. Analysis of Key Findings for Emergency Management

### *4.3.1. Overall Performance*

The meetings and visits with government, Chilean Red Cross, fire department, and individuals allowed the Emergency Management Team to come away with a strong feeling that the efforts of individuals and the community was a major contributing factor in the successful emergency response. Individuals knew what to do in the tsunami-impacted areas. In all locations, people assisted others with rescue and first aid, as well as care and shelter.

The initial response of the central government was problematic. Delayed decisions, miscommunication, communication failures all contributed to a serious impact on the success if the initial response. The government declining initial offers of assistance from other nations delayed some response efforts and support. Despite initial difficulties, the government ramped up and overcame these issues within a few days; they were then able to supply basic services and temporary shelter to those affected by the event. The creation of the Emergency Committee by the new president appears to have enhanced the response as well as begun to address longer term recovery issues. The use of *Mediasaguas* and innovative housing repair and rebuilding programs also were successful.

The Chilean Red Cross response was a strength area. Although their disaster role has traditionally been limited in Chile, and they were not well integrated into the government disaster response plans and efforts, Chilean Red Cross volunteers were among the first to assist people in the impacted areas. Their response involved a wide range of activities, including: the provision of health services, food, water, clothing, tents, and sanitation support services.

Local fire departments and the police in the impacted cities and towns appear to have worked heroically and effectively, even with limited resources, personnel, and initially no outside support.

### *4.3.2. Perception of Readiness*

Prior to the earthquake, officials at all levels and even survivors in the impact area held the belief that since Chile had survived the largest earthquake in recorded history in 1960, the country would be prepared for any subsequent earthquake that may strike. The breadth and scope of the February 27th earthquake and subsequent tsunami, as well as perceived weaknesses about the national and local response to the disaster, were an unpleasant surprise to many.

Because there is frequently an interval of several decades between these large events, much of the hard-earned emergency management experience gained by officials at all levels will be lost or forgotten unless the lessons learned are captured and translated into new laws, plans, and procedures.

While the "culture of preparedness" and public education programs clearly contributed to a lower than expected death toll, the loss of lives among visitors to the coast, particularly in Constitución, reveals that tsunami education and disaster preparedness, in general, must expand to all Chileans. The Chilean Red Cross, local ONEMI officials, and other organizations that teach response readiness must incorporate the lessons learned from this event to ensure that the culture of preparedness is maintained and enhanced.

### *4.3.3. Different Impacts of Earthquake and Tsunami*

The response and recovery issues faced by survivors and government officials in the tsunami-affected areas varied from those in areas solely affected by the earthquake. The destroyed or damaged residences in the inland areas still had available land on site or nearby where the inhabitants could subsist. In most cases, some or all of their possessions could be recovered from the structures.

The tsunami not only destroyed or damaged structures, but also made the immediate area uninhabitable or inaccessible for the survivors who wished to return. Furthermore, the water carried away or ruined many of the possessions that the survivors had left behind (see fig. 12).

### *4.3.4. Differing Impacts in Urban and Rural Areas*

The response and recovery issues faced by survivors and government officials in the affected urban areas varied from those in rural areas.

The size and complexity of the problems facing the fire chief in the large, urban metropolis of Concepción were much different from the problems for the fire chief in the relatively rural province of Cauquenes.

### *4.3.5. Communication Systems*

The communication systems that were in place before the earthquake were not built to a standard that would withstand an event of this magnitude. Many of the radio systems were built with a single point of control. Although plans were developed to work around the failures, by using amateur radio operators to relay information, systems will need to be developed for such events.

Cellular telephone systems were reported to have failed for two hours to days after the event. Many of the cell systems became overloaded. As Emergency Management utilizes cellular telephone for daily emergency management events, planning for such failures and having alternate systems available is a priority.

### *4.3.6. Emergency Planning*

The emergency plans in Chile lacked the necessary detail, scope, redundancy, and flexibility required to effectively and efficiently respond to an event of this magnitude. The Chilean national disaster plan does not specify the roles and responsibilities of the federal agencies in the event of a disaster. The plan implies, but does not state, that the national agencies will act at the direction of the President once information as to the scale and the scope of the disaster is received from the affected regions.

Figure 12. Earthquake and tsunami damage in the coastal city of Constitución.

The national plan outlines in detail the process by which the provinces and regions will report the affects of the disaster to the national government. The February 27 earthquake severed almost all communication between the capital and the affected areas, so an important component of the plan was crippled from the onset of the disaster.

### *4.3.7. Little Local/State/Federal Friction*

The fact that the regional *Intendante* and the provincial Governors are appointed by the President of the Republic resulted in less local/state/federal friction than is customary in the United States. Once the national government mobilized their response, and the considerable resources of the military were brought to bear, the local communities began to receive outside assistance. In some communities, however, this assistance did not arrive until five days after the event.

There was some evidence of friction and potential duplication of response/recovery efforts between the Chilean Red Cross and local government. The friction was evident in the provision of disaster services by the Chilean Red Cross and local government in the areas of field hospital setup and staffing, small grants to disaster survivors and disaster supply provision.

The problems were attributed to the lack of experience in working together and the lack of clear guidelines regarding Chilean Red Cross and local government functions in a disaster.

### *4.3.8. Community Response: Both Adaptive and Maladaptive*

Consistent with the social science disaster response literature, the Emergency Management Team learned that the communities in Chile impacted by the earthquake and tsunami were an asset in the disaster and contributed to the overall response by providing much of the initial search and rescue; sharing food, water and other necessities with neighbors and the larger community; transporting the injured to hospitals before the arrival of organized response agencies; putting out fires at their origins; and providing moral support and encouragement to those who have experienced significant losses. Unfortunately, some members of the community took advantage of the absence of social control to take things from damaged stores, pharmacies, and liquor stores that were beyond the necessities that are reasonable considering the privations caused by the disaster. Looters set some fires, such as the one that destroyed the supermarket pictured in figure 13.

Figure 13. A supermarket in Concepción, which was destroyed in a fire set by looters.

Much of what has historically been described as looting by formal agencies of disaster response and the media are instances in which survivors of the disaster enter damaged grocery stores and pharmacies and take food, water, diapers, and other necessities—items that would be purchased if the stores were open for business. In some cases, business owners, themselves survivors of the natural disaster, are viewed with suspicion when they reenter their damaged facilities to assess damage to the structure and their inventories. In other instances, those who are not members of organized search and rescue teams but are spontaneous volunteers attempting to assist those trapped in the debris of collapsed buildings may be mistaken for looters. Looting has become associated with disasters as a basic assumption or stereotypical image mainly by agencies of social control and the media. The media, when reporting on a disaster, sometimes express surprise when looting does not happen, further perpetuating the assumption that it typically does occur.

There may be instances when looting does occur, often in situations in which a segment of the population (for example, an ethnic group or a social class) harbors grievances based on actual or perceived injustices. This collective sense of injustice may be expressed in looting behavior in situations that validate group perceptions such as the exoneration of white police officers in the Rodney King beating trial that set off rioting that included looting in Los Angeles in 1993. Though looting is rare in the aftermath of natural disasters, looting may occur as an expression of historical or long-standing grievances by groups that have experienced discrimination, abuse, or economic hardship.

#### *4.3.9. Widespread Use of Propane for Cooking*

The widespread use of propane for cooking, in both rural and urban areas of Chile, alleviated many of the mass feeding issues that might be faced in California. The government's mass care feeding plan, with an emphasis on the preparation and distribution of food boxes to individuals and families who cooked their own meals, was very successful. Relatives and friends who brought food into the impact area augmented this plan. Private catering companies also assisted in the effort on their own initiative.

As Chile transitions and modernizes, this feeding model may not function in a future event. Newer apartment buildings, for example, have the propane centralized in large tanks, with the fuel distributed through pipes to the individual units. As this "modernization" of the fuel supply spreads through the urban areas of the country, food boxes may not be the sole solution for mass feeding.

#### *4.3.10. This Event Was Not Chile's Worst--Case Scenario*

An earthquake focused on the metropolitan Santiago area is Chile's worst-case scenario and the one to which their new emergency plans should be addressed. A common refrain stated by Hurricane Katrina survivors in southern Mississippi was, "The water never got this high during Camille." For the inhabitants of the region Hurricane Camille, a Category 5 storm that struck the region in 1969, became the benchmark "worst-case scenario" against which they prepared their emergency management plans. The government of Chile and the Chilean people must not use the February 2010 earthquake as the standard for their preparations for the next event. In any future event that strikes the Santiago metropolitan area, the issues facing the local officials will be like those encountered in Concepción, only amplified by the greater population in the capital.

#### *4.3.11. Temporary and Long--Term Housing*

The new national government moved quickly to implement a temporary housing solution and has developed creative strategies to address the long-term housing problem. Displaced persons are being given two options: stay on their property, or move to an *Aldea*, or temporary housing village. For both options, displaced persons were relocated to *Mediasaguas*, or temporary housing structures. We visited two such *Aldeas*, with each housing between 250 and 500 people. These structures are meant to be temporary. Other options for long-term housing have been developed. Grants for tools and supplies have been distributed to those working on repairing their homes.

## 4.4. Recommendations for California

The lessons learned from the Chilean earthquake are not unique; each has been a lesson learned from most if not all preceding major disasters. These included: communications failures, inadequate plans, inadequate pre-disaster coordination and slow or poor initial leadership by government officials. Each of these issues clearly has implications for California. Although none of these areas are new, each should be freshly analyzed and compared with existing plans, communication systems, and coordination in California. Each applicable lesson should be addressed in current catastrophic planning efforts underway in California, as well as future plans and exercises.

1) The initial response and resilience of individuals and communities is something we can learn from. Expanding and enhancing efforts in California to make individuals, communities, towns, and cities better able to survive the first few days of a catastrophic event is critical. The American Red Cross (ARC) and other NGOs are vital to this capability in California. The already strong ARC partnership with state and local governments should be maintained and enhanced when possible. Since it may take several days for outside assistance to arrive following a catastrophic event, this point cannot be overemphasized.
2) Communication system failures limited the ability of a central government to assist impacted communities and NGOs to coordinate their mass care resources. This limited the ability of the government to communicate tsunami warnings and other information. It also delayed the response, since the government did not know (in some case for several days) the impact and needs of local governments. California needs to develop, maintain, enhance, and regularly test primary and back---up communications systems.
3) It appears that Chile believed it was prepared for a large earthquake. They had lived through a *M*9.5 quake in 1960 and felt they were ready for another large quake. Their plans were not adequate for the effects of this earthquake. They also lacked backup plans when primary plans failed. The implication of this for California reinforces what we already know. Plans need to be scalable and flexible to meet all hazards and all sizes of events. Backup and redundancy need to be incorporated into overall plans, and Continuity of Operations/Continuity of Government (COOP/COG) plans need to be regularly exercised.
4) Unique methods were required to meet the care and shelter needs of the displaced population. Although plans for congregate care shelters did exist, very few were used. People instead stayed in their damaged dwellings (sometimes due to fear of being looted), stayed with family, or traveled to other parts of the country. Delivery of food, water, and the maintenance of sanitation had to revolve around the large number of people sheltering in place. This needs to be further analyzed, as it may need to be reflected in plans in California. American Red Cross and local communities are prepared to support people in congregate care shelters. More recent catastrophic planning has begun to address

the provision of support to people sheltering in place in damaged homes. These efforts may need to be enhanced.

5) The looting that occurred in Chile was a surprise to response officials as well as social scientists but both public and private sector organizations, including NGOs, must consider security for damaged businesses as a priority in California's multihazard planning. Clearly, the class and ethnic divisions that become acute in specific cases of actual or perceived injustice may also emerge in natural disasters, particularly in California's urban areas, which have historically experienced unrest. While it is unlikely that looting will be overlooked in disaster planning, Chile's experience merits a reaffirmation of the need to rapidly respond in securing damaged businesses and critical infrastructure.

# 5.0. HEALTH SERVICES

## 5.1. Introduction

The primary goal of the Health and Medical Services Team was to understand the medical disaster response strategies and operations of Chilean agencies for the earthquake on February 27, 2010, including:

- Perceived or actual failures in disaster preparation that impacted the medical disaster response;
- Post---disaster health and medical interventions to save lives and limit suffering; and
- Lessons learned by public health and medical personnel as a result of their experiences.

Despite devastating damage to the health care and civic infrastructure, the health care response to the Chilean earthquake appeared successful due to several factors:

1) The ability and resourcefulness of the medical community to respond without centralized control in the early response phase
2) The dedication of the health care community to maintain patient care under austere conditions

3) The capacity to rapidly re---establish utilities and transportation infrastructure that supported health care in most areas
4) The rapid mobilization of national and international resources
5) The expectation of the Health and Medical Services Team, and t he Red Cross delegation as a whole, was that many of these lessons would be transferable to California's disaster planning and response.

## 5.2. Methodology

The Health and Medical Services Team conducted interviews in Chile with members of the Ministry of Health, Pan American Health Organization, Chilean Red Cross, and hospital directors in Santiago and the regions that suffered the most damage from the earthquake. Interviews were scheduled with individuals identified as responsible for disaster response planning and/or operations within the Ministry or health care institutions. In-person interviews lasted between 60 and 90 minutes. Personnel from national, regional, and local levels of the health care system were represented, as well as the national response agency and national and regional Red Cross.

The investigators conducted semi-structured interviews. Although the team did not employ a validated outline or tool, the interviews explored common themes, including:

- The respondent's professional responsibilities for disaster planning and response and tenure in their current position (this had particular relevance due to the change in government 10 days after the earthquake)
- Pertinent emergency response planning related to written plans, implementation of drills, and provision of training conducted by the institutions or outside sources
- The respondent's actual experiences during the disaster and how these compared to prior planning for communications, management of resources, patient care and evacuation, injury epidemiology, and factors related to resiliency of the health care system and recovery activities.

The Health and Medical Services Team also asked respondents to describe the strategies to prepare for a future disaster event. Lastly, respondents were invited to review their system disaster response, identifying specific strengths and concerns related to future disaster preparation.

Interview notes were transcribed into Word files to be later used for content analysis. Investigators verified impressions for accuracy. Themes and data were then organized into a final report.

## 5.3. Chilean Medical Organizational Structure

The Ministry of Health of Chile (Ministerio de Salud de Chile, also known as MINSAL) is a cabinet-level administrative office in charge of planning, directing, coordinating, executing, controlling, and informing the public health policies formulated by the President of Chile. The ministry is divided into two sub-components: (1) public health and (2) health care. An Undersecretary directs each component. The Ministry of Health oversees the health care delivery system of clinics and hospitals, as well as the public (FONASA) and private (ISASPRE) insurance systems. One other division is responsible for purchasing pharmaceuticals and equipment. The country is divided into nine administrative regions; each has a Director who reports to the national Ministry of Health. The regions are further subdivided to form a total of 28 districts. The Health Ministry supports and manages 180 public hospitals and 3,000 primary care health clinics across the country. The majority of health care (for approximately 80% of the Chilean population) is provided through this public system.

MINSAL coordinates funding, supplies, administration, transportation, and bed control. The Ministry also coordinates special programs such as nursing or convalescent homes, dialysis programs and a home ventilator program. In 2009, in response to the H1N1 pandemic, the Ministry started a computerized bed control program in order to coordinate bed utilization between the largest public hospital facilities and private hospitals. The system posts available critical care beds in both sectors, prioritizing the least expensive beds.

There are also private hospitals/clinics in Chile that are regulated by the Ministry of Health. These are concentrated in the larger population centers of

Santiago, Valparaiso, and Concepción and provide healthcare to a more affluent segment of the society. The private sector has no internal organization and does not coordinate emergency preparedness planning with the public sector. After the earthquake, MINSAL used beds in the private sector facilities to accommodate patient transfers from impacted public hospitals. There was precedence for this, since MINSAL routinely leases beds from the private sector for use by public sector patients for surge capacity during the winter respiratory illness season.

MINSAL includes a Chief of Preparedness and Response and a Director for Hospital Recovery—a new position since the earthquake. Each regional director is responsible for coordinating preparedness and response within their jurisdiction.

## 5.4. Disaster Response Overview

Disaster response in Chile is managed through a centralized federal agency (ONEMI), which could be considered the equivalent of FEMA at the national level and the California Emergency Management Agency (CalEMA) at the California state level, that coordinates response across various agencies and among federal, regional, and local jurisdictional levels in Chile. The Ministry of Health (MINSAL) is responsible for health and medical response, which is an Essential Support Function (ESF 8) in the United States National Response Plan (REF). International assistance is supported and coordinated by the Pan-American Health Organization (PAHO), the regional agency of the World Health Organization.

### *5.4.1. Disaster Plans*

The team was consistently told that emergency response plans were inadequate and failed at all jurisdictional levels for the response to the February 27th earthquake in Chile. Even federal officials and those inside the agency readily admit that ONEMI response to this incident was inadequate. Within the health sector, regional and local representatives stated their plans were inadequate and never "taken off the shelf," since the plans did not deal with a disaster of this magnitude. Planned response was further disrupted because many emergency response facilities were destroyed and unusable; for example, in Talca, the main fire station, Red Cross headquarters and police station were all on the coast and rendered inoperable by the tsunami. The regional emergency operation center for the Health Ministry in Concepción was severely damaged, and all their communication equipment was destroyed or inaccessible, including satellite phones, computers, and radios.

### *5.4.2. Command and Control*

Initially, there were no communications throughout Chile. During the first 24 hours there was very little situational awareness and the local responders and hospitals were, for the most part, on their own. Therefore, command and control was initially based at the local level but expanded to include federal and regional levels of support and coordination as situational awareness and communications improved. To develop situational awareness, the central Ministry of Health sent representatives to the regions on the first day after the quake. Situational awareness gradually increased over the second and third days, but coordinated situation status in one region was not fully achieved for nearly 10 days.

The individual medical center/hospital directors had a large degree of autonomy in managing the medical response in their institutions and communities. Once communications were reestablished, situation status reports and resource requests went from the local level to both the regional level and directly to the federal Ministry in Santiago. Over time, federal and regional offices of the Ministry of Health coordinated federal resources and resolved specific issues raised at the local level, such as support for dialysis units, patient transfers, and allocation of resources such as field hospitals and volunteers.

### *5.4.3. New Government*

Especially challenging was the change of administration in Chile 10 days after the earthquake, resulting in new personnel at the federal and regional levels. This provided an opportunity for the new government to modify the response structure. For example, an Emergency Committee was formed and assigned to provide direct executive command and control over the response.

Given all these challenges, the response at all levels was surprisingly rapid, resourceful, and successful.

### *5.4.4. Pertinent Timeline for Response Coordination*

Day 1: characterized by a chaotic local response with no communications and many response capabilities, offices, and equipment destroyed.

Days 2-3: characterized by local response only with no outside supplies.

- Initial organization of services on site with likely shortages of equipment or supplies.
- Communications improved but minimal overall situation status.

Days 4-5: characterized by improving communications, situation status, and organization;

- Outside resources begin to arrive

Day 6 and beyond: significant outside resources and help arrive to support local responders.

- Begin to establish field hospitals.

### *5.4.5. Pertinent Statistics*

- Approximately 80% of Chile's population of 16 million persons was located in the earthquake impact area.
- There were 131 hospitals in the southern five Chilean regions most directly affected by the quake. This is approximately 70% of the hospitals in Chile. Portions of 59 hospitals received irreparable damage, an additional 16 hospitals were judged unusable. By some estimates, more than 4,000 hospital beds were lost (of which 70% were occupied at the time of the earthquake), requiring relocation of thousands of hospitalized patients.
- 17 field or temporary hospitals were erected between one and three weeks after the earthquake.
- Private hospitals were not significantly impacted because they are centered primarily in Santiago and Valparaiso, with some in Concepción.
- There are approximately 3,000 primary care centers in Chile providing outpatient services, and most were intact post---disaster.
- To date, there are 521 confirmed deaths and 56 persons missing; half of the deaths were due to the tsunami and most others due to collapse of adobe/brick style houses.
- No statistics are available regarding increased mortality due to evacuation of hospitals, loss of power, or other impacts on the health care system.

## 5.5. Immediate and Early Response

### *5.5.1. Instinctive Response*

In the absence of organized, functional response plans, the actual response to the Chilean earthquake by many individuals was based on "instinct." The team was repeatedly told that Chile has a "seismic culture" in which everyone knows what to do in an earthquake. The common lore is that after an earthquake strong enough to throw one down, one should run out of the building when the shaking stops (and go to high ground if near the coast).

Initial search and rescue efforts were also instinctual and done by first responders (Fire) and by members of the community (see fig. 14). In one community hard hit by earthquake and tsunami, all victims found alive were rescued during the first 24 hours, before any outside assistance was available.

Figure 14. Search and rescue efforts atop the collapsed Alto Rio apartment building in Concepción. (Photo courtesy of Jorge Arturo Borbar Cisternas).

The most notable example of the instinctive response in the health sector was the rapid evacuation of hospitals. Two medical directors from different hospitals went to their hospitals immediately following the quake, arriving in less than 30 minutes to find the facilities fully evacuated. This was accomplished because the staff and patients did not wait for orders from any centralized command structure or the person in charge, but immediately helped patients out of the building. Patients helped other patients out. Due to the debris in corridors, it was impossible in most cases to wheel patients in beds or wheelchairs, so they were carried or walked out. One hospital the team visited had a full census of 140 patients at the time of evacuation, but none in critical care. The other evacuated hospital in Santiago had 275 patients on seven floors, including critical care areas. After two days in a nearby parking structure, some patients were placed in non-damaged parts of the hospital and about 200 patients were transferred to other hospitals or discharged home.

### *5.5.2. Hospital Evacuation and Engineering Evaluation*

Many, if not most hospitals in the two most highly impacted regions (Bio-Bio and Maule) were initially evacuated due to structural and non-structural damage. After structural evaluation, one-third of all public hospital beds were lost in these two regions. In Santiago, only one hospital was seriously damaged and evacuated. Evacuation of patients from the impacted areas was used to decrease health care demand in those regions.

Initially, damaged hospitals evacuated all patients immediately following the earthquake. Decisions to evacuate were made by doctors and other staff on duty in the middle of night. They could not assess the true extent of damage, but fallen ceiling tiles and wall panels, broken windows, and cracks in walls were widespread in many facilities. Subsequently, many patients were relocated to undamaged or minimally damaged parts of the hospital.

Engineers responded locally, and others were dispatched by the Ministry of Health days after the earthquake from Santiago. Decisions to move patients back to relatively undamaged areas of the hospital were likely made prior to official seismic engineering evaluation. On the other hand, one official observed that many hospitals were deemed destroyed, when in fact they were determined to be repairable after more definitive engineer evaluation; initial assessments may have, in some instances, been completed with the intent of replacing old or outdated facilities and equipment.

### *5.5.3. Utilities*

Power was restored in some areas within hours and in most areas within seven days. Generally, hospitals were not evacuated solely for lack of utilities. The team was told of only one hospital that evacuated due to lack of utilities, but patients were repatriated within about 24 hours. Hospitals used power generators that were on site or provided by the Pan American Health

Organization or by local emergency services. Ministry of Health coordinated with responders and power companies to ensure that hospitals received priority for electrical power.

There was widespread primary disruption to the water delivery system in the impacted regions. It was reported that several hospitals had wells that continued an uninterrupted supply of water to specific facilities. Other hospitals received water transported by truck and placed in large emergency water tanks or bladders for storage. No one was aware of written plans for rationing a limited supply of water, but priorities were given to patient care and staff hygiene, supplemented by alcohol hand sanitizer. PAHO, Ministry of Health, and Red Cross all provided consultation for water treatment and evaluation of water supplies.

### *5.5.4. Patient Evacuation*

The Ministry of Health estimates about 2,000-3,000 patients were transferred between hospitals; however there are no figures available yet for the number and type of patients evacuated or their geographic distribution. Very few patients were evacuated by air (rotary wing). Most planned evacuations were done by ambulance, within 3-7 days after the earthquake, and consisted of previously hospitalized patients. Fortunately, most major traffic routes in the country remained passable.

Locals related that many patients were transported in any available vehicle, including military transport, trucks, and private vehicles, which is consistent with the limited numbers of ambulances in the regions and the prolonged transport times; in optimal conditions, it is six hours from Concepción to Santiago. Some patients were taken to nearby hospitals, only to find them incapacitated, and had to go hours further to unaffected hospitals in other regions. Normally, inter-hospital transfers are coordinated through a web-based bed tracking system run at the central Ministry in Santiago. As Internet services were re-established within days after the earthquake, this system was used to coordinate; however, the team was unable to determine if evacuations were coordinated early on and how destinations were determined, considering the variable road and hospital conditions.

### *5.5.5. Surge Capacity*

Surge capacity is the ability to provide care to accommodate a sudden increase in the number of patients. Surge capacity involves hospital beds, but also includes staff, supplies, and the decision-making process. Consistent with the general sense of failure of planning for such a catastrophic event, no one interviewed described following specific pre-event plans for how to manage the increased patient load; however, many components of normal procedures and emergency planning were used.

The Ministry of Health claims that there was little health care surge, either inpatient or outpatient, following this disaster. Local facilities, however, reported an outpatient surge; one unofficial estimate from the Red Cross was approximately 60,000 (mostly ambulatory) injuries that resulted from the earthquake.

#### 5.5.5.1. Beds

Procedures utilizing a web-based database existed for coordinating bed control and patient transfer (see section 5.5.2, Hospital Evacuation and Engineering Evaluation, above) but were not initially available due to disruption of power and communications. Decisions to use intact or lightly damaged portions of hospitals and the strong inclination to maintain and treat patients within the community greatly helped maintain existing capacity, despite large losses in bed availability. Evacuation out of the impacted regions was used to relieve pressure on the health care system, but data regarding numbers of patients transferred are not available. The private care system provided surge capacity in unaffected regions through leasing beds to the public sector. By days 5-7, some field hospitals were established, eventually numbering 17. About half of the field hospitals were supplied from international donors, while others came from the Chilean military and Red Cross. Many are still in use and being replaced by temporary hospitals, consisting of modular trailer-like or container units expected to be in use for several years while the hospitals are rebuilt.

#### 5.5.5.2. Staff

Since the earthquake struck at 3:34 a.m. Saturday morning, staffing in most facilities was at a minimum. Many staff reported to their work sites soon after the earthquake and others reported at the morning change of shift, after being assured of the safety of their family. With many houses destroyed or heavily damaged (nearly one-third of those in the Maule region), some staff were unable to report for duty. Tragically, the looting that occurred in the Concepción area in the first few days before security was established by the military resulted in more staff staying home to protect their household and family. Experience of PAHO in Latin America, and in this earthquake, are that most health care workers stay on the job or report for work. Because most medical personnel are employed by the public health care system, they are able to move and practice freely between hospitals during an emergency; in many cases, staff was directed to other facilities where the need was greater.

#### 5.5.5.3. Volunteers

The volunteer response was early and strong. Volunteers from the community showed up at the hospitals, but many had no health care skills, so were of limited use. Other medical volunteers were sent from Santiago and other parts of the country. The Ministry of Health set up a volunteer system for this disaster with web registration. The Chilean Red Cross plays a major role in supplying volunteers for health care facilities and providing direct health care. Their response took several days to become coordinated, but subsequently mobilized significant numbers of volunteers.

International response to the earthquake was strong, but assistance was initially declined by Chile until the scope of the disaster and needs were fully appreciated. Spanish volunteers, including medical specialists and psychologists arrived within 72 hours. After seeing images on television, an Argentinean group traveled overland to provide care at a local community hospital. Another hospital received a Japanese surgical team on day four. Other medical teams arrived from Russia, Cuba, Spain, and Italy. International response was coordinated through the federal government and PAHO/WHO. Except for mental health, the need for volunteer medical personnel was of limited duration.

#### 5.5.5.4. Military

The Chilean military played a key role in supporting many aspects of the response, including health care and patient movement, but were not immediately mobilized. The military supplied, erected, and maintained some of the 17 temporary field hospitals in the impacted locations. There was initial reluctance to deploy the military, but their support and performance generated uniform praise.

#### 5.5.5.5. Supplies

Initially, hospitals had to depend on existing supplies. One health care provider described the fear during repeated trips into the severely damaged building during the first few days to gather supplies. However, a lack of supplies was not described as one of the major problems following the earthquake, perhaps due to local and regional stores of supplies provided regularly by the national Ministry of Health. Food was one commodity initially in short supply for the community and for hospitals, and was exacerbated by the initial looting of stores in Concepción. Resource requests after the earthquake were made to the regional and central Ministry. Medical and food supplies from Santiago began arriving on days 3-4 into most of the impacted areas in the southern region. Additional supplies such as vaccines for Hepatitis A and H1N1 were sent from Santiago. In some areas, donated supplies became a problem, especially medications that did not meet Chilean requirements like labeling in Spanish. Private companies are not integrated into the national response plan but donated goods and supplies spontaneously.

#### 5.5.5.6. Crisis Care

Crisis care refers to care under austere conditions with resource shortages that requires a change from the usual standards of care and necessitates allocation of resources based on providing the most care to the most people. This is not a term or concept that the team encountered in Chile. No one spoke of legal concerns or ethical dilemmas of care that are persistent and pervasive concerns in the United States. Much of the health care resiliency in this catastrophic situation can be attributed to the initiative of local facilities and emergency responders and to the flexibility of medical staff to provide care under disaster conditions.

Another likely factor for the Chilean resiliency was acceptance of the austere conditions as the only alternative after this disaster. Hospitals, especially older facilities in small towns, suffered considerable damage, resulting in patient care being provided in the parking lots or under very crowded conditions in the limited safe areas and ongoing treatment in tent-style field hospitals in their community. Health officials are confident that everyone received needed care, although occasionally at facilities some distance from their communities, and no one was denied care.

The surge of ambulatory patients following the earthquake was handled wherever usable space was available, often outdoors. The primary care system, which is part of the public health system, was largely intact following the earthquake and helped to provide ambulatory care at local outpatient clinics. Local health officials and responders describe a large surge of injured patients, but agree that most could be handled locally and that the majority of the cases were considered minor trauma.

According to PAHO, after the first few days of the emergency, the number of individuals seeking outpatient care actually decreased and was less than normal for the time of year. This was attributed to the population focusing on other priorities, such as cleaning up earthquake debris and confirming the well-being of family members in affected areas, as well as the desire on the part of some to stay close to their damaged homes for fear of losing property to looting.

### *5.5.6. Hospital and Health Care Continuity and Response*

Hospital care was most impacted due to both structural and nonstructural damage that made the facilities, or portions of facilities, unusable (see fig. 15). The Ministry of Health indicated that the emergency medical system and primary care system remained intact throughout the disaster. Only twelve primary care centers, also run by the Ministry of Health, were significantly damaged. During the first week, the emphasis of the central Ministry of Health was on in-patient care, but by the second week the need was more for outpatient care. Field hospitals saw primarily new patients, since many patients who were in the hospital when the earthquake occurred had been transferred or discharged by the time field hospitals were established. The local facilities saw a major surge of ambulatory injuries in the first few days following the disaster.

Figure 15. Damage at Felix Bulnes Hospital, outside of Santaigo. As of July 2010, the hospital was still partially closed due to unrepaired damages.

#### 5.5.6.1. Medical Record Keeping

In some hospitals, medical records fell from open shelves to the floor, in disarray. Few hospitals in the impacted regions had electronic medical records (EMR), although that is being gradually implemented. However, EMR also would have been problematic due to the initial lack of power and Internet access. In the early period after the earthquake, care was performed under austere conditions, and thorough documentation of medical records was not a priority. It is not yet known what clinical information was captured or whether medical records for those first hours are available.

#### 5.5.6.2. Special Medical Needs

Chile has fewer nursing homes per capita than the United States. Presumably, the elderly and those with chronic health conditions remain in the family home. There are no data available from central sources of the impact to nursing homes or evacuation or transfer of nursing home patients due to earthquake damage. None of the interviewees noted this as an issue, even when specifically asked. The team was also unable to observe or obtain specific information regarding people with functional and access needs, such as those who are wheelchair bound or have other mobility issues.

The home ventilator program supports approximately 1,000 patients across Chile. Public health nurses, who are normally responsible to monitor the care and status of ventilator patients, were dispatched to each residence to ensure their safety or have them moved to hospitals with generator power when necessary. MINSAL was not aware of any mortality among the home ventilator patients, even with the reported widespread power outages.

The southern region around Concepción has less than 1,000 outpatient dialysis patients. Local health centers reviewed their status, and special efforts were made to supply power and water to the clinics. Some patients were transported to Santiago if their local centers could not accommodate them. It is not known whether there was any morbidity or mortality among the dialysis-dependent population or how many patients required transfer to Santiago.

#### 5.5.6.3. Behavioral Health

One of the major health care surge needs was for mental health support and counseling. Medical staff, first responders, and volunteers were affected, as well as the general population. Initial demand was driven by anxiety and grief induced by the disaster, but the need is expected to persist for a long period of time. This need was and continues to be met through a combination of volunteers, especially from NGOs and coordinated by MINSAL and PAHO.

- The psychiatric hospital in Valparaiso was severely damaged by earthquake and most of the patients were released to the community for outpatient care, since there are few mental health inpatient facilities in Chile.
- The Ministry of Health helped identify patients receiving mental health services on an outpatient basis and contacted the individuals to make sure they had their necessary medications.
- In Talcahuano, due to the severity of earthquake and tsunami damage and the large number of fatalities, mental health issues were identified early as an issue. Group therapy sessions were started in shelter camps to counsel and support victims.
- The national Comité de Emergencia is supporting a program currently focused on 4,500 families residing in temporary houses in the "villages." All the residents lost their homes, but some also lost jobs and family members. The Comité has coordinated with representatives from NGOs to live within the communities and communicate to the residents information regarding how the government can help them, that is, finding jobs/education. The goal is to focus on resiliency and recovery so that the "community" won't become a permanent client.

### *5.5.7. Communication and Media*

After the earthquake, all communications in the affected southern regions were interrupted. Communication between the south and the central government in Santiago was also impacted, as was local communication briefly within Santiago itself. The only intact communication was via hand-held emergency radio at the local level. Communications were restored over days, but the timeframe was highly variable, depending on modality and location. In Concepción, text messaging became available within the first day, followed by cell phone communication and Internet. By the end of the first day, half of the hospitals in the Bio Bio region had been contacted through text messaging. Although the central Ministry of Health had radio communication capability, it functioned inconsistently. The Ministry offices were able to contact Concepción and Talcahuano two days after the event. As soon as possible, the Health Ministry in Santiago began videoconferences with affected zones. One particularly devastated community on the coast had only local radio communication for 4 days.

The media played a mixed role. Only a single radio station remained on the air in Concepción immediately following the earthquake, but it was very helpful since no one had situational awareness or means of receiving information. Several people commented that some of the media, especially television, exacerbated the chaotic situation by reporting only on the worst of the disaster, rather than helpful emergency messaging and risk communications.

### *5.5.8. Morbidity & Mortality*

The official death toll at the time of the visit was 521. Approximately half died as a result of the tsunami and half from the earthquake. Earthquake deaths were nearly all from burial under collapsed adobe structures. Health care statistics and epidemiology of injuries are not yet available. Health officials consistently made several unanticipated assertions:

- There were very few serious traumatic injuries resulting from the earthquake inside or outside of damaged hospitals.
- Of the patients who were transferred to remote facilities, most had been hospitalized at the time of the earthquake.
- Surgeries immediately following the earthquake were due to the usual surgical problems, rather than from earthquake---related injuries.
- There were no major fires, so there was a lack of burn patients, contrary to anticipated results of an earthquake in California (see Section 3.4).
- No one died due to mass evacuations from hospitals, numerous patient transports over long distances, or loss of power and utilities within the communities. Even with careful questioning, there was no reported loss of life for patients with special medical needs, such as dialysis and ventilator patients.
- Post-disaster, there were few cases (and no major outbreaks) of intestinal or respiratory diseases.

The team did hear a few reports of earthquake related mortality including a head-injured patient from the earthquake who died during transport out of the area and four people who arrived at a damaged hospital with myocardial infarctions and all died. One person related that two or three patients died due to collapse of hospital ceilings or walls, but the Ministry of Health did not validate this.

It is remarkable that the Chilean health care system was able to maintain operations and care for existing and new patients, with little apparent morbidity from the disruption to hospital care and long transports.

## 5.6. Health Services Key Findings and Recommendations

The following represent conclusions drawn from the information presented from meetings and interviews that have implications for health services preparedness and response in California. Limitations of drawing conclusions are that these data are from unstructured interviews and represent a perspective influenced by the individual's job, jurisdictional level, and experience following the disaster. One challenge in extrapolating these Chilean experiences to planning and response for a large California earthquake is the inability to determine the relative roles and impact of infrastructure, organizational development, and cultural differences on our conclusions.

1) **Evacuation**. Hospitals should plan for evacuation and develop a process and criteria for evacuating. No recommendation can be made on the advisability of encouraging spontaneous evacuation, as occurred in Chile. That question requires input from seismic and structural experts. It is unlikely that rapid spontaneous evacuation would be initiated in California; more likely, staff would wait for an order to evacuate and patients would wait to be assisted. After a significant California earthquake, it must be assumed that even hospitals that are structurally sound will have significant non-structural damage such as supplies and some equipment on the floor, fallen ceiling tiles, and other superficial damage, but this should not be the sole reason to evacuate. In a catastrophic event with widespread damage, health care facilities may initially evacuate, but should abandon the facility only when absolutely unable to provide care or when engineers deem the structure unsafe to occupy.
2) **Crisis Care**. After the Chilean earthquake, care was provided under austere conditions in many health care facilities and within the community. Hospitals in Chile initially operated in parking lots, and then made early decisions to use safe, functional areas of hospitals and clinics, which allowed them to continue improvised care despite difficult conditions. This could be the situation after a catastrophic earthquake in any country. Even hospitals that are structurally sound will be operating under austere conditions due to damage to equipment, lack of power and water, and no communications or Internet. Care will be provided outdoors or in improvised locations

due to evacuations and influx of new cases without the capacity to accommodate them in emergency department or usual care locations. Certain resources and materials will rapidly be depleted or unavailable. Few physicians in the U.S. have experience under austere conditions; some will quickly adapt while others will require strong leadership to be effective. Although not needed in Chile, guidelines and plans to support these crisis conditions of care should be available in California, as recommended by the Institute of Medicine (IOM), the Agency for Healthcare Research and Quality (AHRQ) and several professional organizations. In Chile, cultural factors and the flexibility of health care professionals and of the population likely contributed significantly to the resiliency of the health care system.

3) **Field Hospitals.** Field hospitals were used to add health care bed capacity to the damaged Chilean hospital system. Their experience demonstrates the advantage of placing these field hospitals on the site of existing health care facilities to extend capacity at sites where the public normally seeks health care and allowing patients to remain in their community. This arrangement also augments structurally safe facilities or usable portions of facilities and allows access to undamaged equipment and supplies. Chile is replacing tent hospitals with temporary modular hospitals that can be used until permanent facilities are rebuilt.
4) **Utilities.** Hospitals must be prepared to be completely self-sufficient, including sources for power and water, for up three days minimum, five days preferably, and one week optimally. Joint Commission standards currently recommend hospital self-sufficiency for 72 hours, and this is also consistent with current and new guidelines from AHRQ. Hospitals should have more than one generator and have technical specifications ready to order replacements. Public emergency services should plan for the capacity to provide required power, water, and food after that time. Emergency deployment of hospital generators in Chile was a problem, because authorities had not anticipated the special hook-ups and secured pads for mounting that were required but not available at various facilities. This resulted in some generators not being utilized and a delay in the use of others.
5) **Epidemiology.** Assumptions regarding morbidity following a catastrophic earthquake in California are very different from the Chilean experience and suggest the need for further assessment. Lack of fires and resulting serious burn injuries may be attributed to

infrastructure differences, but it is more difficult to explain the few numbers of seriously injured in Chile. Conclusions are premature until epidemiologic analysis of injuries and illness is available.

6) **Health Care Worker Staffing.** The reports of staff shortages were not quantified, but appeared far less than some planning scenario estimates in the United States. Earthquakes do not create the same fear factor for personal safety in health care settings as does an infectious disease event; however, key issues are the ability of staff to travel to work and the need to ensure safety of their families. Chileans have strong local family and community support which allowed healthcare personnel to be comfortable in leaving their homes and reporting to work.

   There will be many committed health care workers in the United States, but they will need major support from outside responders. Emergency plans should include a process to accommodate mutual aid consisting of medical teams and facilities from outside the impacted area. Medical personnel need the ability to move and practice freely between hospitals within the disaster region or the state during an emergency. This emphasizes the importance of California Disaster Healthcare Volunteer systems and other volunteer agencies, but this ability also needs to be extended through interstate emergency assistance such as that coordinated by the Emergency Management Assistance Compact (EMAC) and the Medical Reserve Corps (MRC).

7) **Health Care Surge.** Field hospitals are rationally situated at existing hospital sites to provide medical treatment at locations where the public is accustomed to seeking care. Since field hospitals and experienced disaster health care workers are available in limited quantities, serious consideration should be given to accepting international aid after a catastrophic earthquake in California. Coordination of this aid is the responsibility of the United States Department of Health and Human Services, but is likely to be mired in political and bureaucratic barriers.

   Bed control and patient tracking systems are critical, especially when widespread evacuations and patient transfers are necessary. Chile has such a system that was utilized as soon as Internet access and local administrative organization was re-established. California is still developing a statewide bed control systems and does not currently have uniform patient tracking.

8) **Volunteers**. While volunteers are critical, their skills need to be matched with needs, whether basic volunteers or skilled medical teams. It is important and necessary to pre-validate and determine the skills of international teams, as well as domestic and in-state teams.
9) **Security.** Looting caused increased security concerns for personal property, impacting availability of health care staff; this emphasizes the need for immediate security measures in the community and at health care facilities.
10) **Behavioral Health**. Behavioral health needs occur early and are persistent through recovery. Disaster mental health planning and surge capacity is critical.
11) **Communication.** Although most communications were re-established rapidly in Chile, initially it was very frustrating and difficult to have no communications. Health and other critical agencies and institutions need redundant communication systems to communicate situation status and resource requests within and between jurisdictional levels. Initial situation status may be impossible to determine without functional communications, which in turn makes resource allocation decisions very difficult. Emergency hand-held radios proved particularly useful for local primary responders, allowing them to coordinate initial activities.
12) **Media**. The sensationalism of media can be detrimental to initial response efforts and to public stress. Proactive and competent risk communication is critical in order to have any control over the public message.

## 5.7. Questions for the American Red Cross that Arise from the Chilean Earthquake and Response

Many of the issues noted in this section could have significant consequence for the Red Cross in its preparation and response to a California catastrophe. Based on the Chilean experience, these are some of the healthcare-related questions the Red Cross should consider in its catastrophic disaster planning and development of partner relationships.

1) Is the Red Cross prepared to deliver post-disaster services in settings where there is no water, communication, sewage, or electricity?
2) Does the Red Cross have the capacity to deal with ongoing post-disaster mental health needs for clients and workers, especially if mass casualties occur?
3) Can the Red Cross address worker and volunteer issues, such as long and stressful work days and worker illness or injury attributable to fatigue, stress, or worry about family and loved ones?
4) How can the Red Cross better prepare to support clients who are unwilling to leave their damaged residences or who prefer non-traditional shelter settings such as tents and cars?
5) Does the Red Cross have adequate equipment, healthcare supplies, and partnerships to support the hygiene, medical, and chronic healthcare issues, including functional and access needs, for clients in long-term mass care environments?
6) Can the Red Cross obtain timely, accurate information and communicate to clients and the general community the availability of local health and medical resources, the status of healthcare facilities, and locations for obtaining triage or primary care services?
7) Basic community first aid provided at local service delivery sites, emergency aid stations, or through outreach teams in impacted neighborhoods may become an important post-disaster Red Cross function, especially if hospitals sustain widespread damage or experience patient surge due to disaster-related major injuries. Is the Red Cross able to provide personnel and first aid supplies to accommodate and deliver first aid services at the level that may be requested?

# 6.0. Volunteer Management

## 6.1. Introduction

The Red Cross Volunteer Management Team found that Chilean Red Cross faced many challenges during their response, including limited disaster response experience, lack of volunteer management coordination, limited disaster response preparedness, lack of a central national data base and little communication. Moreover, the earthquake and tsunami affected many of the buildings that housed the Red Cross, other volunteer organizations and emergency response agencies. Even with these challenges, they provided a successful response within communities, due to a pervasive culture of volunteerism and commitment to one's community.

Sheltering in Chile is conducted by ONEMI (equivalent of FEMA). The Chilean Red Cross and various religious non-profit organizations serve in a support role for sheltering. There was a very limited amount of traditional sheltering provided in Chile, as most of the population stayed on their own property to protect their belongings and assist their neighbors. Building *Mediasaguas* (single family wooden structure houses), helped to support the housing needs of many individuals and families.

## 6.2. Methodology

We met with volunteers from many of the affected regions. We were taken on a tour of some of the hardest hit areas, which allowed us to talk first hand with Chilean Red Cross volunteers who were affected by the disaster. The Chilean Red Cross has a very limited role within the government structure at this time. In the past, the Chilean Red Cross's role has been limited to health and safety education and did not include disaster response. It is the hope and intention of the Chilean Red Cross to build stronger relations with Government entities in advance of future disasters. ONEMI Director, Vicente Nuñez, recognizes that Chilean Red Cross has an important role in disaster planning. He advised that ONEMI is in the process of reorganizing, and that the Chilean Red Cross will be involved in future disaster-planning discussions (oral communication, 2010).

## 6.3. General Chilean Red Cross Volunteer Background

Chilean Red Cross has approximately 2,500 volunteers nationally and received 30,000 spontaneous volunteer inquiries after the earthquake/tsunami. The Chilean Red Cross has a predominantly older volunteer base, of which 97% is female. In the past, young adults were not well accepted, but that seems to be changing. There is, however, a large gap in experience and situational knowledge.

Chilean Red Cross has a youth program, intended to better integrate the youth into the volunteer system. Youth work in over 200 schools across the Chile. They act as school ambassadors, providing first aid training and sex/HIV education. Volunteers younger than 18 are not allowed to respond to disasters due to liability; however, when Chilean Red Cross volunteers turn 18, they can train other new young volunteers.

As presented to us by Chilean Red Cross leadership, the major focuses for the Chilean Red Cross are:

1) Youth
2) Risk Management (disaster services)
3) Health
4) Social Activities & Wellbeing (elderly, poor, immigrants etc)

Chile did not plan for a disaster of this magnitude. Though Chile has previously gone through one of the largest recorded earthquakes, it was not near any major cities, in a sparsely populated part of the country. Nonetheless, many Chileans felt that they were prepared to deal with any large earthquake.

## 6.4. Timeline of Chilean Red Cross Response

National Chilean Red Cross members and other response agencies met with ONEMI within the first few hours after the earthquake. Chilean Red Cross participated in ONEMI meetings for the first three days after the earthquake. They realized that this disaster had surpassed their response capacity, so they decided to step back from the national response to focus on local community response (Oral communication, Nelson Hernandez, CRC Director of Risk Management, 2010).

Red Cross Chapters handled the initial response with local volunteers. This continued for five days before they received mutual aid volunteers, who were sent by the Chilean Red Cross's headquarters (Oral communication, Manuel Quezada, CRC Executive Director, Bio-Bio Region).

## 6.5. Chilean Red Cross Volunteer Commitment

Chilean Red Cross has very few paid employees. Paid staff mainly consists of executive directors (National and Regional Directors), secretaries, and janitorial staff. All other personnel are volunteers, who are expected to work a minimum of 20+ hours per week, after required training is complete.

## 6.6. Chilean Red Cross Volunteer Activation

Volunteers are instructed to report to their Chapter within 3 hours for large disasters and 12 hours for small disasters (Oral communication, Cristian Winkler Santander, Regional Director of Talca Region CRC. The Chilean Red Cross and other volunteer agencies advised us that it is understood that volunteers should report to their local office following a disaster. In Concepción, many volunteers were unable to report because of the Red Cross building being inaccessible or damaged. In most cases, an alternative location was not designated before the event.

All government EMS workers in Concepción had been trained to report to one predetermined meeting place in the center of the city. The meeting place was heavily impacted and inaccessible. The local Fire Department is now aware that a 'Plan-B' should have been included in their emergency plan. As a result, they have created a new disaster plan with three potential meeting places. This also highlights the importance of selecting meeting sites that would likely remain accessible after disaster.

## 6.7. Chilean Red Cross Volunteer Identification During a Disaster

The Chilean Red Cross does not provide ID badges to authorized Red Cross volunteers. During a disaster, Red Cross vests signify who is a Chilean Red Cross volunteer. This raises concerns of identifying who is a legitimate Red Cross volunteer. This system also does not distinguish between trained Red Cross workers and untrained spontaneous volunteers. The Chilean Red Cross has recently implemented new protocol for 'uniforms,' which consist of a new vest, a shirt, and cargo pants. At this time, there does not appear to be a need to create ID badges or implement background checks for personnel. This is an additional indicator of the difference between the Chilean Red Cross and American Red Cross response during a disaster. The American Red Cross has a clearly identified need for ID badges, as do most other entities that respond and work together during disaster in the United States.

## 6.8. Chilean Red Cross Volunteer Database

Each Region keeps it own volunteer database. When speaking with Chilean Red Cross personnel, many concerns were raised regarding a non-centralized database. If an area is adversely affected during a disaster, there could be a loss of data regarding names of volunteers, training records, etc. There is no standard to decide what information is kept regarding volunteers. There are numerous databases with multiple formats, and each region collects and stores its volunteer information differently. This can result in challenges in identifying what skills are available in each region, as well as in determining the quantities and types of resources available for deployment.

Not having a centralized database aligns with the Chilean focus on local community empowerment, rather than solely relying on National leadership. Chilean Red Cross Chapters are able to keep customized records according to their specific needs. Smaller databases are easier to manage, which is an essential quality for an organization run mostly by volunteers.

## 6.9. Chilean Red Cross Retention

The Chilean Red Cross is challenged with retaining youth volunteers after they leave school due to the requirement of volunteering 20 hours per week. Young adults often get jobs or go to university, and do not have 20 hours of time to donate. This is an unfortunate downfall of the 20-hour expectation, which otherwise helps to maintain an active core group of volunteers. Youth should be considered the future of the Chilean Red Cross adult volunteer base, and yet there are few programs in place designed to boost youth retention.

## 6.10. Communication

Communications systems throughout all affected regions were impacted, if not destroyed, during the earthquake and tsunami. Many fire stations were badly damaged and communications between response organizations were (mostly) not working.

Some response messages were delivered through the media. In Concepción, the media focused on the Alto Rio building (fig. 4) and did not portray any other damage. As a result, many emergency services rushed to that building, and many other affected areas did not receive the immediate support they needed.

These communications breakdowns demonstrate the importance of backup communications systems in all emergency plans. Specific disaster protocol, such as sending all damage reports to a few central locations, are essential so that emergency services do not have to rely on media to know where to go.

## 6.11. Fire Department Volunteer Management

The Chilean Fire Dept consists of 311 fire stations. It is the oldest volunteer organization in Chile. Even the National *Bomberos* President, Miguel Reyes Nuñez, is a full-time volunteer. Chilean Fire Dept has 38,000 *Bomberos*, 4,000 of whom are women (see fig. 16).

Figure 16. American Red Cross Delegates meet with National Bomberos President, Miguel Reyes Nuñez.

Mutual aid is prevalent within the fire department. They request mutual aid by proximity— they first request aid within their district, and then ask outside their district for help. This is very similar to the mutual aid and deployment process that ARC uses.

Volunteer *Bomberos* receive subsidies, medical insurance, and life insurance if harmed in the line of duty. Dispatchers and drivers are the only paid staff and are on duty at a dispatch center or the fire station. Volunteers are not always in the stations (due to personal commitments, jobs, and so forth) and often respond directly from home to meet the fire truck at the incident. Firemen are primarily alerted by siren and cell phones.

There is no minimum volunteer time commitment; however Miguel Reyes Nuñez did not feel that they were often 'short-staffed'. This shows the sense of commitment, accountability, and responsibility amongst the volunteers.

*Bomberos* pay a monthly fee to be volunteers. This sparks an interesting debate regarding the increased sense of value if someone pays to be part of something; this is worth considering further as a point of debate. For example, some argue that if someone pays a fee to be a member, they tend to participate more than a person who is a member for free.

The fire department works with youth aged 12-17. This program encourages those older than 18 to become *Bomberos*. The success of this initiative shows the importance of working with youth. This program has been a positive influence on communities, and has led to closeness with the community and propagated a culture of volunteerism. This should be considered a lesson on how to change the culture of future communities through youth programs.

## 6.12. Volunteer Management Findings and Recommendations

Volunteers in Chile tend to be very resilient. They are able to work effectively with little or no direction from National Headquarters. This has been part of the reason that they were able to respond immediately to such a devastating event.

The following are recommendations for improvement in California, based on observations in Chile:

1. There need to be redundant ways of communicating with personnel.

- Ensure that radio communication equipment is in place and that personnel are trained to use radio equipment. Write a plan and procedures for using radio communications during disaster response.
- Utilize automated call out systems using an out-of-state server to ensure that the server is not damaged. The automated call out system

should allow activation of calls by telephone or online. The automated call out system should send messages by phone and text to ensure that the message is sent as soon as possible.

- Utilize social media for personnel who would want to subscribe to social media. Although this was not prevalent in Chile, news reports from Haiti showed that social media messages were active before cell phone and landline messages were working.
- Have a backup plan in place so that personnel know what to do when all communication is down. For example, Red Cross Volunteers could be advised to report to their local office for instructions, and designated personnel would be trained to open offices, coordinate volunteers, and use radio equipment.

2. Emergency plans need to be flexible and include alternative options in case primary plans are unable to be executed.

- It is imperative to have at least one backup alternative plan and to communicate disaster plans with personnel. For example, there should always be more than one meeting place and more than one designated person fully trained in each position in case the primary person is not available.
- Plan for technology not working; ensure that staff has access to manual volunteer databases and the paperwork needed to handle staffing.

3. All volunteer leadership, at all levels, need to know emergency plans.

- All responders at all levels should be aware of the disaster plan.
- Leadership should be fully trained in their role and strongly aware of the roles and responsibilities of additional team leaders and members.

4. Exercises need to be done on a regular basis, with volunteers rotating in their normal roles.

- If only a few volunteers know the plan and they are unavailable after a disaster, no one will know what procedures to follow.

5. Involve local officials in regional planning.

- Plans should be developed in coordination with government entities.
- Partnerships with written Memoranda of Understanding (MOUs) should be developed in advance with non-government organizations.
- Inter-agency training is strongly recommended for entities that would work together during a disaster.

6. Perform a realistic assessment of life-essential systems, such as potable water sources and emergency medicine supplies. Preparedness on the part of the community is a key element in dealing with these issues.

- Personnel should be trained for the probability that core services may not be available and know how to respond and support the community in such conditions.

In summary, both the Red Cross in Chile and in the United States seem approachable to local residents, and have strong community relationships. The success of the Red Cross will depend on community relationships and public impression of the Red Cross to ensure that communities feel that they can approach the Red Cross for aid. It is crucial that American Red Cross continue to maintain and develop a strong relationship with government and non-government entities to ensure a unified and effective response.

# 7.0. Executive Management

## 7.1. Introduction

How a government responds after a disaster usually captures the headlines. But most often, it is the role that government plays in preparing for these types of events that can be the single biggest factor in minimizing not only the event's initial toll, but also the recovery time necessary to bring a community back to a healthy, functional state. In this regard, an examination of how the government in Chile responded during the immediate aftermath of the earthquake and related tsunamis is appropriate. The relationship between the government and the local Red Cross is also important to understand the social dynamics of the recovery process. In addition, some observations about the plans and programs the government is now instituting for the recovery phase may offer insight into where things may evolve in terms of capabilities, culture, and priorities.

Information compiled for this report includes direct observation, interviews with various government, non-government, and community officials, data received directly from each of these groups, and external data that is referenced to provide context and foundation. In each case, attribution is cited and footnoted.

## 7.2. Analysis of Social Interactions

### *7.2.1. Social Priorities*

It is very clear that in the United States, a high priority and expectation is placed on government and recognized organizations such as the Red Cross to keep people safe and provide for their basic human needs during times of emergencies and disasters. This expectation was reinforced by the many reports that emerged after Hurricane Katrina. Criticism of the U.S. Government's response ranged from Nancy Pelosi's report (California Policy Desk, 2006) entitled "Katrina Response: a Scandal of Incompetence and Cronyism" to the Select Bipartisan Committee Report to Congress entitled "A Failure of Initiative"(House of Representatives, 2006), Despite the fact that this was the single largest disaster to ever strike the United States in terms of total losses, total costs, and total resources required, it is clear that the expectation remains among some in this country that response to such events, however cataclysmic, will be timely, effective, and efficient.

In Chile, it appeared that central government believed that the expectations of the Chilean public were much lower than those of the American public. In our meeting with the head of the Emergency Committee that had been commissioned by the President to oversee the response and long term recovery efforts, there was an admission that because the country had not suffered a large disaster in many years, this had led to a gradual reduction of planning, exercising, and investment in preparedness infrastructure, thus contributing to the deterioration (and ultimate failure) of their systems and other emergency response and recovery capabilities (Oral communication, Cristóbal Ibáñez, Director, President's Emergency Committee. July 20, 2010). Both Chilean Government and Red Cross officials told the delegation that there had been surprise at the level of public criticism to the government's response. Both agreed that the positive public perception of the Red Cross had contributed to the government's early efforts to engage the Red Cross in more meaningful and proactive ways, including participation on the President's Emergency Committee.

Chile has an ongoing, albeit innate public-private relationship in that most of the 100% volunteer fire fighters come from the private sector. The new president, who comes from the private sector, called immediately upon the private sector to reinvent the country's emergency management organization.

### *7.2.2. Relationships between Agencies*

From our discussions with officials at the national level down to local government and individual responder agencies, there was little to suggest that the various organizations had worked closely with each other during planning and preparedness activities within recent times (Oral communication, ONEMI, Chilean Red Cross, Regional and Local Government Officials, July 2010). In 2006, the Board of Governors of the Chilean Red Cross strengthened the disaster charge of the national society, but progress has since been slow in co-opting national and regional government to reconsider this broader role (Oral communication, Chilean Red Cross officials, July 2010). A pattern that emerged from our various meetings was that each organization with whom we met seemed quite clear about their own role and responsibilities for preparedness, response and recovery activities, while their understanding and expectations of different levels of government or outside organizations starkly differed.

While everyone with whom we spoke saw the Red Cross in a positive light, there were in many cases very low expectations about the ability of the Red Cross to play a significant role in disaster services. In certain settings, that reduced expectation may have accurately reflected the presence and capabilities of the local Red Cross Chapter. However, in at least one case, there appeared to be more local Red Cross capability than that which local government was initially aware of or perhaps preferred to acknowledge. In this particular region, the Red Cross was poised to begin providing financial relief (approximately $300 per family) and requested records from the Mayor's office regarding the affected families in order to establish and confirm the need for assistance. The mayor requested that his office be given the funding so they could distribute the assistance. When told that was not possible and that the Red Cross would need to work directly with the clients, a decision was reached to not provide the Red Cross with the requested records. Red Cross officials told us the rationale they were given by the Mayor's Office was that it would have been harmful to the public's perception of the government's response if people saw that they were receiving assistance more from the Red Cross than they were from their government.

During a visit to one of the tsunami-affected areas in this region, we were pleased to see brand new mattresses being distributed to the housing units in the relief camp there. The local Mayor met with us inside the camp and provided us with his talking points and priorities for what he felt his area needed from their national government. It was only later that we learned from the local Red Cross officials that the mattresses we saw being distributed had been sitting for weeks in a local warehouse, and that it was very likely the presence of the Study Team that finally prompted their distribution during our time in the relief camp (see fig. 17).

As for its relationship with the general public, Red Cross has a strong reputation across Chile. It is volunteer led, has a local presence all over Chile, and is seen as part of the overall global Red Cross movement. Many Chileans are vague on what exactly the Red Cross provides in Chile, but they almost all have a very good overall impression.

Figure 17. Mattress delivery to *Mediasaguas* 'village' outside of Tumbes.

### *7.2.3. Responsibilities for Documentation*

A key finding for most of the team members was the paucity of data relating to the earthquake response. Even four months after the event, many areas of statistical information had yet to be developed or completed. When we met with the National Fire Chief in Santiago and inquired as to certain fire-related reports, he told the study team that even basic fire incident reports in Santiago for the days immediately following the earthquake had yet to be completed and compiled (Oral communication, Miguel Reyes Nuñez, National Chief of Bomberos. July 27, 2010). The inconsistency with which data were being mined and developed was reflected by all disciplines, even hospitals. By contrast, medical records, systems, and documentation are traditionally held to a high standard of precision compared to many other industries (American National Standards Institute, 2010).

The general lack of emphasis on data collection and information development led to national statistics that were inconsistent and inconclusive. While there is general agreement about the key areas of failure surrounding the country's response to the earthquake, it was never evident that these were developed as a result of any process of statistical collection and analysis, but rather through a combination of anecdotal feedback and written or verbal comparisons of common problems encountered. Without the development of additional means to identify or confirm specific incident response data, both analysis of past events and planning for future ones will remain challenging. It may be somewhat telling that while there was a general acknowledgement that better systems for collecting and analyzing data were necessary, this was never cited as one of the leading problems or areas of future concern by any of the organizations with whom the Executive Team met.

### *7.2.4. Culture*

Our team noted an amazing resiliency on the part of the Chilean people. Many of the affected population, even in the wake of this enormous disaster, chose to remain completely self-reliant. If people lived near the coast, they knew to evacuate for higher grounds after any earthquake where they had fallen down or could not walk due to the shaking. In almost all cases, these were not lessons learned from national or local government officials, nor even from the Red Cross. Most natives have learned what to expect and what they should do from their parents, grandparents, and others who have passed the information down through generations of people, many who experienced similar events during their lifetimes. We learned about many people who had returned to their properties as soon as the tsunami threat had passed, only to find their homes either nearly destroyed or, in many cases, missing completely. Despite this, many opted to remain at their land and manage their own recovery, rather than seek the assistance that was being organized and available at various relief camps.

At the national level, a history and culture of self-reliance also continues to be strongly evident. During the initial hours following the earthquake, then-President Michelle Bachelet refused early offers of international aid, including both formal and informal offers from the United Nations (oral communication, Enrique Ganuza, United Nations Resident Coordinator. July 26, 2010). .Critics have claimed this as a contributing factor to the delays seen in meeting the basic needs of victims in the first days following the disaster.

Coming on the heels of the Haiti earthquake, the timing of the Chilean earthquake coincided with when many aid agencies began struggling to raise money for relief efforts, amid fears that the world community was becoming "disaster weary." Within weeks, the world had new disasters to which their attention had been diverted, including the Gulf oil spill. The longterm financial implications for Chile are daunting, with a total recovery cost that has been projected at $30 billion.

## 7.3. Implications for California and the American Red Cross

### *7.3.1. Relationships*

Our findings reinforced the importance of our "pre-disaster" relationships with all of our governmental, non-governmental and community partners. Properly done, these relationships require an organizational commitment, as well as a significant investment of time and personnel. The number and complexity of these relationships will vary based on local nuances, but as a general rule it is vital to ensure Red Cross representation and active participation in disaster policy, planning, response, and recovery activities at all levels appropriate to the specific setting. It was no surprise to learn that the areas in Chile that made the most effective use of the Red Cross and its resources were the very areas where some level of interaction had been ongoing before the earthquake struck.

In the United States, our government's approach to disaster response is well defined and emphasizes a "bottom up" approach, where local government is considered as having primary responsibility for emergency management. The National Response Framework (Department of Homeland Security, 2008) describes the tiered approach and flow in this manner:

> "Even when a community is overwhelmed by an incident, there is still a core, sovereign responsibility to be exercised at this local level, with unique response obligations to coordinate with State, Federal, and

> private--sector support teams. Each organization or level of government therefore has an imperative to fund and execute its own core emergency management responsibilities."

In any particular setting, therefore, the appropriate connections must be made at all relevant levels. Paul Schulz, the CEO for the Los Angeles Region, met with the ARC Delegation as well as with Chilean national government and Red Cross officials during the final three days of our visit. The summary of his observations included the following:

> "When there is a major disaster in Los Angeles, the natural political leader will be the Mayor of Los Angeles, just as it was the President in Chile. It is critical that the Red Cross, and the regional CEO in particular, have a personal relationship with the Mayor in order to be a key part of the ad hoc kitchen cabinet that will inevitably be formed. Mayor Riordan formed this ad hoc group during the Northridge earthquake, and most likely this would happen again. As the Mayor will reach out to those he or she trusts during this crisis, having a personal working relationship is a prerequisite for being on this senior team."

### *7.3.2. Social Priorities*

There is little to suggest that Americans are willing to accept less readiness in this country, despite the significant costs associated with maintaining such a state of preparedness. While the American Red Cross plays a vital role in the nation's readiness infrastructure, it is ultimately the responsibility of local government and local communities to ensure that mass care equipment, supplies, and facilities have been made available to provide for the needs of the community during the immediate post-event phase of a response. As we saw in Chile and on many occasions in the United States, the early days after a large-scale disaster will best reflect and will ultimately serve as a barometer for local community readiness.

American Red Cross representatives should rightfully consider themselves as advocates and spokespersons for ensuring that mass care needs within their community are properly represented. This can assist local emergency management officials in developing an all-inclusive approach to emergency management in their respective jurisdictions that considers all of the Emergency Support Functions listed in the National Response Framework.

Michael Kleiner, Director of Emergency and Disaster Response for the Los Angeles Region of the American Red Cross, believes that many communities have yet to achieve balance with respect to their planning and approach to community preparedness:

> "A comprehensive emergency management program must consider all of the Emergency Support Functions outlined in the National Response Framework. Like any chain of related activities, we will only be as good as our weakest link when actual events stress and test our systems. If, for example, we haven't properly planned and resourced a community's mass care needs, we are doomed to keep repeating similar mistakes as we have seen in the past during large disasters, where people begin sensing early that they may need to fend for themselves. This gives rise to unrest and activities such as looting, which immediately requires the diversion of precious and scarce emergency resources away from the initial disaster. In our community, Mass Care was recently rated by the various jurisdictions in our county as our largest collective capability gap, yet only a small and disproportionate fraction of the funding from numerous emergency preparedness grants was earmarked to address this shortfall. We need to do better at balancing the actual needs with our planning and preparedness priorities."

### *7.3.3. Responsibilities for Documentation*

Most members of the study team found the lack of data to be initially somewhat disorienting, particularly for the scientists and clinicians on the team who find themselves immersed daily in information relating to their professional roles. Despite the lack of data, there was agreement among the study team that many of the best lessons learned were about things that are inherently difficult to "measure," such as the cultural distinctions noted in this report.

Dr. Richard Hinrichs, Chief Officer of Disaster Services for the San Diego Red Cross and former Federal Coordinating Officer in FEMA Region IX, helped put things in perspective regarding the lack of hard data:

> "We have often in the past realized only after a disaster certain things we should be tracking better. Nearly every after--action plan you will ever see will discuss the need for better communication and recommendations on ways to better gather and use real--time information. Despite our best efforts, we all know from experience that many of our most significant learning moments come from intuition and observation, rather than from a data set. Albert Einstein once said, "Not everything that can be counted counts, and not everything that counts can be counted." This statement exemplifies many of the most significant lessons we learned during our time in Chile, such as

our observations about the ability of Chileans to react to the needs in front of them, taking action without waiting for some central authority to tell them what to do."

### *7.3.4. Culture*

Unfortunately, the individual mindset in the United States has not yet evolved to include a culture of personal preparedness. Many people in this country look to government and organizations such as the American Red Cross to be prepared to address the needs of communities and disaster victims, regardless of past attempts to shift some of that responsibility back to individuals and families.

The American Red Cross has long advocated personal responsibility. The Red Cross provides Community Disaster Education classes, sells disaster preparedness kits and supplies, and works in many settings with local government to continue efforts to shift our culture towards a greater sense of personal responsibility. A key for this may well be the consistency in messaging. As the public continues to be barraged with advice on what exactly constitutes adequate personal preparedness, subtle differences in this messaging may be hurting our collective efforts.

Craig Fugate, Director of FEMA, has cited improved personal preparedness as having the largest potential for improving our nation's level of readiness. Over the past several decades, this theme of personal responsibility for disaster readiness has been widely espoused, and during that time our culture has continued to reject this responsibility, not so much by words as through a collective failure to act. Fundamentally, something must change or we will likely continue to expend significant amounts of time and money promoting a concept that has not succeeded in making itself a part of our national culture. One fundamental change could be to strictly limit the ambiguity of our messaging by jointly working to develop one common script. Accomplishing this on a national level would require a national forum, and should be encouraged, but if even local communities that share a media market can agree on a singular message, the promotion of that singular message by many organizations would almost certainly be more conducive to achieving the desired changes in social behavior than would result from multiple and often contradictory messages being delivered by multiple organizations.

Ellis Stanley, former President of the International Association of Emergency Managers, previously worked as the Emergency Manager for two large metropolitan areas of the country (Atlanta-Fulton County and the City of Los Angeles). He currently serves on the Homeland Security Advisory

Council's Emergency Response Senior Advisory Committee. Ellis has long advocated the notion of a simple, singular, standardized message for the public:

> "People have often told us they end up confused as a result of the different preparedness messages they hear from various sources within their communities. For whatever reason, many organizations feel the need to "brand" the message by developing their own unique catch phrases or marketing strategy. Then, when people hear inconsistent and contradictory messages about how they should prepare, how much food and water is adequate, and how long they can expect to wait for help, it becomes that much harder to convince them to take action. If we can develop a National Response Framework, why can't we agree on a National Preparedness Framework that could provide the same level of standardization regarding the necessary elements of readiness, including a common message for the public that has been vetted and agreed upon by the emergency management community?"

From a financial and fundraising perspective, past experience both here and internationally has shown that the window of time available to effectively fundraise for disasters is unreliable at best. Uncontrollable variables such as other disaster events occurring simultaneously or shortly afterwards can greatly limit the time frames for generating adequate financial support for our events. This reinforces the importance of fundraising as one of the basic things we must all focus on early during disaster relief operations.

# 8.0. RECOMMENDATIONS FOR CALIFORNIA

Throughout each section, we have presented findings and recommendations that will be useful for professionals in specific areas. In this section, we highlight overarching themes in Chile's response that resonated through multiple disciplines and areas of response. These represent the main strengths and weaknesses that surfaced during the event and could be put to use in preparing California for potentially catastrophic events.

**1. Conducting comprehensive exercises (including joint Government, private sector, NGO, emergency responder, and community exercises) before the event is paramount to surviving and thriving after a natural disaster.**

**Justification**. When we visited the coastal towns hardest hit by the tsunami, we were told that communities often drill at least twice a year, practicing full evacuations to higher ground. Emergency drills need to be done on a regular bases with key players actively exercising in their specific roles. It is important to exercise within one's own organization; however, it is strongly advisable for entities to work together when drilling.

**2. Individual resiliency and effective networking with local partners are vital to the continued success of a community after a disaster. It is critical to empower people to be prepared.**

**Justification.** Most of the hardest hit areas in Chile were cut off from aid and communication with the capitol for several days following the earthquake. All of the responders in these areas cited their personal resourcefulness and local partnerships (for example, between the firefighters, police, emergency management, and the Red Cross) as critical in their ability to help their communities in the difficult first few days. The already strong ARC partnership with state and local governments should be maintained and enhanced when possible.

**3. Education about what will happen during the event is an important part of preparing for a disaster; this information can save lives during a disaster.**

**Justification.** Most sources in Chile cited their "culture of resiliency" as a central factor in their ability to respond. It is clear that earthquake drills are more common than in California and that earthquakes are a more visible part of their society. At the same time, lack of information about the true nature of great earthquakes and tsunamis appears to have contributed to injuries and fatalities. For example, many people were taught to head for high ground if shaking was so strong they could not stand; this appears to have saved many lives in the tsunami. At the same time, it is not clear that the occurrence of multiple waves over many hours in tsunamis was anticipated.

**4. Emergency and earthquake professionals should work with representatives of the print and broadcast media before the disaster to determine how to best serve the community.**

**Justification.** The media played a mixed role. Only a single radio station remained on the air in Concepción immediately following the earthquake, but that was very helpful since no one else had situational awareness or means of receiving information. At the same time, several people commented that some of the media, especially television, exacerbated the chaotic situation by reporting only on the worst of the disaster, as well as emphasizing the dangers from looters and the potential shortages of supplies.

**5. Emergency plans need to be redundant, flexible, and detailed to handle the unexpected in very large disasters.**

**Justification.** Emergency plans in Chile were in place for the Chilean organizations we met with, but in all cases they were described as inadequate for the situation they faced. Deficits were especially seen in flexibility and alternate plans. For example, plans had only one meeting place that could not be accessed, or one designated person in each position who was injured or out of town.

**6. Recognize the competing personal and professional demands that will be made on an organization's staff after a disaster and include this in emergency plans.**

**Justification.** Immediately after the Chilean earthquake and tsunami, many critical staff members stayed with their property or left work to see if their families were safe. Other staff simply could not get to work. Staffing shortages have the potential to hamper response efforts in the hours, days, and weeks following a major disaster.

**7. Organizations need to plan for nonstructural damage and the potential need to evacuate even without structural damage.**

**Justification.** After a significant California earthquake, it must be assumed that even buildings that are structurally sound will have significant non-structural damage with supplies and some equipment on the floor, fallen ceiling tiles, and other superficial damage, but this should not be the sole reason to evacuate.

**8. California must recognize vulnerabilities in our communications systems and make comprehensive backup plans to avoid complete communication collapse. It is important to do this as individual organizations, but also to team up with other organizations.**

**Justification.** Although most communications were re-established rapidly in Chile, initially it was very frustrating and difficult to have no communications. Health and other critical agencies and institutions need redundant communication systems to communicate situation status and resource requests within and between jurisdictional levels. Initial situation status may be impossible to determine without functional communications, which in turn makes resource allocation decisions very difficult. Emergency hand-held radios proved particularly useful for local primary responders, allowing them to coordinate initial activities.

**9. Explore mechanisms to encourage building owners to adhere rigorously to existing building codes.**

**Justification.** In Chile, buildings performed extremely well, due to strong, well-enforced building codes. By law, if there is building damage or injuries within a building, the building owner may be liable. Both building professionals and lay people in Chile reported that this law serves as extra incentive for building owners to adhere to the building code during construction.

**10. Collect all possible data about each disaster when it happens.**

**Justification.** Many of the consequences of the Chilean earthquake and tsunami have not yet been quantified, such as the numbers of fires or injuries caused by the events. Each disaster is a unique opportunity to learn how society is affected by the events and this information if captured can support researchers who are trying to minimize future losses.

# 9.0. Recommendations for the American Red Cross

**1. The ARC (American Red Cross) should actively seek to create, develop and maintain close working relationships at all levels of local and state government that play roles in disaster preparedness, disaster education, disaster response, and recovery from disasters.**

**Justification**. The health of our pre-disaster relationships will be key in whether the Red Cross will play a comprehensive role in decision-making at the highest operational levels of an event. Our ability to do so will not only enhance our services to clients, but can provide a mechanism for the consideration of certain cost-containment strategies that may help influence efficiencies.

**2. The ARC should promote the standardization of preparedness messages provided to the public.**

**Justification.** As the preeminent national organization and one widely recognized as a leader in the area of disaster preparedness, the ARC should attempt to play a larger and more central role in the process of standardizing the many preparedness messages that routinely bombard the general public currently. As a national organization, ARC is uniquely suited both geographically and by area of expertise to take a greater leadership position in the pursuit of this goal.

**3. As a participating response agency, the ARC should play a greater role in the development of research in order to contribute to the knowledge gained by such endeavors.**

**Justification.** The absence of data for the Chilean events was profound by our standards in the United States. This further highlighted the importance of not only data collection, but also the need to establish methods to process and analyze such data in order to provide meaningful information for purposes of better informed planning and mitigation. The ARC and all other response organizations with significant disaster roles should contribute to this effort beyond the simple provision of data.

**4. The ARC should anticipate the significant gap in Disaster Health Services and Disaster Mental Health personnel available for a catastrophic event and seek to address this by the pre--arrangement of ancillary licensed personnel.**

**Justification.** The medical capabilities of any community will easily become overwhelmed in the wake of a major earthquake. The Red Cross has licensed personnel who provide Disaster Health Services and Disaster Mental Health. Without pursuing additional strategies, we will not have adequate human resources in these areas. The prearrangement by MOU of Public Health and other medical personnel by Emergency Support Function (ESF) could provide for appropriate percentages of community resources to be dedicated to each of the necessary response functions.

**5. The ARC should develop as a portion of their Disaster Plans a "pre--scripting" of basic initial activities in order to ensure that early actions are organized even in the absence of emergency communications. These scripted activities should be realistic and scalable to accommodate larger scale events.**

**Justification.** In virtually every after-action plan, communications is cited as a key factor contributing to problems during the response phase. Pre-scripting provides a standardized approach to initial organization of response efforts, while scalability should include a strategy for the controlled degradation of staffing ratios and systems. For example, predefined criteria for reduction to more austere levels of staffing and care should be considered in advance of an event, so as to avoid being forced to make such decisions during actual operations.

# 10.0. Sources and Acknowledgments

## 10.1. Sources

American National Standards Institute: accessed August 11, 2010, at http://publicaa.ansi.org/sites/apdl/default.aspx.

Astroza, M., Cabezas, F., Moroni, M., Massone, L., Ruiz, S., Parra, E., Cordero, F., and Mottadelli, A., 2010, Intensidades Sismicas en el Area de Daños del Terremoto del 27 de Febrero de 2010: Universidad de Chile, Santiago (in Spanish), accessed September 28, 2010, at http://www.eqclearinghouse.org/20100227-chile/wp-content/uploads/2010/04/Informe-de-Intensidades-M.-Astroza-y-otros.pdf.

California Policy Desk, 2006, Pelosi: Katrina response a scandal of incompetence and cronyism: California Chronicle, February 8, 2006, accessed August 10, 2010, at http://www.californiachronicle.com/articles/view/5646.

Comité de Emergencia (President's Emergency Committee), Government of Chile.

Department of Homeland Security, 2008, Introduction, *in* National Response Framework: p. 5, http://www.fema.gov/pdf/emergency/nrf/nrf-core.pdf.

GEER (Geoengineering Extreme Events Reconnaissance) Team, 2010, Geo-engineering reconnaissance of the February 27, 2010, Maule, Chile, earthquake: v. 1, April 15, 2010, http://www.geerassociation.org/GEER_Post%20EQ%20Reports/Maule_Chile_2010 /Cover_Chile_2010.html.

U.S. Geological Survey Earthquake Program, 2010a, M8.8, Offshore Bio-Bio, Chile, *in* Worldwide recent earthquake information: accessed September 30, 2010, at http://earthquake.usgs.gov/earthquakes/recenteqsww/Quakes/us2010tfan.php#details.

U.S. Geological Survey Earthquake Program, 2010b, The Modified Mercalli Intensity Scale: accessed July 25, 2010, at http://earthquake.usgs.gov/learn/topics/mercalli.php.

U.S. House of Representatives, 2006, A Failure of initiative: Final Report of the Select Bipartisan Committee to Investigate the Preparation For and Response to Hurricane Katrina, http://www.gpoaccess.gov/katrinareport/mainreport.pdf.

## 10.2. Acknowledgments

We are extremely grateful to the Chilean Red Cross, who hosted our visit and provided much of the information. We particularly acknowledge Lorenza Donoso Oyarce (National President), Francisca Ringger Villarroel (Vice President), Patricio Acosta Sansarricq (Vice President), Yolanda Muñoz Conte (CEO), Andrés Recabarren (Director, Responsabilidad Social Corporativa), Robinson Talavera Cerda (Director Nacional de Salud), Cristian Ramírez Sepúlveda (Punto Focal Agua y Sanieamient), Francisco Cruzat Araya (Director Nacional de Juventud), Jorge Romo (Encargado de Logistica), Maricella Huentemilla Hidalgo (Subdirectora, Logistica), Nelson Hernández Marulanda (Director Nacional de Gestión del Riesgo), Manuel Quezada Melillan (Coordinador Regional de Emergencia, Concepción), Leticia Escamilla (logistical support), Cristian Winkler Santander (Director in Maule Region), and Marcela Winkler Santander (Comunicaciones in Maule Region).

We further benefited from discussions with representatives of the Chilean government, the Director of the Comité de Emergencia de la Presidencia, Cristobal Lira Ibañez; members of ONEMI, especially the Director Victor Nuñez and Johaziel Jamett, Andres Salgado, Jefe de Gabinete, Cristian Torres, and ONEMI representatives in Talca; as well as Jorge Enriquez and Francisco Irarrázaval at the Ministerio de Vivienda. We are grateful to academic colleagues, Dr. Sergio Barrientos at the Universidad de Chile and Dr. Mario Pardo, Dr. Rafael Riddell, Dr. Ernesto Cruz, and Dr. Diane Comte at Universidad Católica de Chile. We also wish to thank the Bomberos Commander Marcos Oyarzo, Cuerpo de Bomberos de Talcahuano and Concepcion, and the National Bomberos Commander, Miguel Reyes Nuñez, and Dr. Celso Bambaren, Ministry of Health, Dr. Luis Suarez (OPS/OMS), Dr. Vladimir Pizarro, Hospital Félix Bulnes, Lic. Gisela Acosta, Jefe Departamento de Emergencias y Desastres—MINSAL, and Dr. Marta Werner Canales. We are grateful to the Los Angeles Consulate of Chile for the guidance offered to those who planned and executed the logistics of the visit, Nancy Carrillo--- Rodriguez and Monica Leon. Finally, we appreciate the logistical support provided by So Young Kim and our interpreters, Pedro Pavez and Megan Codd.

# 11.0. APPENDICES

## 11.1. Appendix A

American Red Cross Multidisciplinary Delegation to Chile—Participant List

| First Name | Last name | Title | Agency | ARC Delegation Team |
|---|---|---|---|---|
| Jim | Goltz | Earthquake and Tsunami Program Manager | California Emergency | Emergency Management |
| Richard | Hinrichs, PhD, CEM | Chief Officer, Disaster Services | American Red Cross, San Diego/Imperial Counties Chapter | Emergency Management |
| Randy | Linthicum | Chief, Disaster & Client Services Bureau (CA DSS Sheltering) | Department of Social Services | Emergency Management |
| Gregory | Smith | Emergency and Disaster Response Director | American Red Cross, Bay Area Chapter | Emergency Management |
| Scott | Underwood | Emergency and Disaster Response Assistant Director, Los Angeles Region | American Red Cross, Los Angeles Region | Emergency Management |
| Michael | Whitehead | State Mass Care Coordinator, State of Florida | Florida Department of Business & Professional Regulation | Emergency Management |
| Michael | Kleiner, CEM | Emergency and Disaster Response Director | American Red Cross, Los Angeles Region | Executive/ Policy |
| Paul | Schulz, MBA | Chief Executive Officer | American Red Cross, Los Angeles Region | Executive/ Policy |

**(Continued)**

| First Name | Last name | Title | Agency | ARC Delegation Team |
|---|---|---|---|---|
| Ellis | Stanley, CEM | Vice President of Dewberry, Western Emergency Services, Emergency Management and Homeland Security Branch / Board Member, ARCLA | Dewberry | Executive/ Policy |
| Howard | Backer, MD, MPH | Associate Secretary for Emergency Preparedness | California Health and Human Services Agency | Health Services |
| Mary | Chasin, RN | Health Services Supervisor | American Red Cross, Los Angeles Region | Health Services |
| Jeff | Upperman, MD | Director of the Trauma Program at Childrens Hospital Los Angeles, and an Associate Professor of surgery at the Keck School of Medicine of the University of Southern California. | Childrens Hospital Los Angeles | Health Services |
| Lucy | Jones, PhD | USGS Seismologist | U.S. Geological Survey | Science/ Engineering |
| Keith | Porter, PhD | Associate Research Professor - Civil, Environmental, and Architectural Engineering | University of Colorado | Science/ Engineering |

| First Name | Last name | Title | Agency | ARC Delegation Team |
|---|---|---|---|---|
| Deborah | Weiser | Geologist, contributor and recorder | U.S. Geological Survey | Science/ Engineering |
| Kelly | Mulock | Emergency and Disaster Response Personnel Manager | American Red Cross, Los Angeles Region | Volunteer Management |
| Nancy | Carrillo-Rodriguez, MSPA | Disaster Response Liaison, Grants & Special Projects | American Red Cross, Los Angeles Region | General Logistics Manager/ Translator/ Consulate Liaison |
| Monica | Leon Uribe, MA | Development Associate | American Red Cross, Los Angeles Region | General Logistics/ Translator/ Consulate Liaison |
| Charlie | Sardou, MBA | Communications, Community Disaster Education and Marketing Director | American Red Cross, Los Angeles Region | Communi-cations |

## 11.2. Appendix B

The table below summarizes each of the meetings we had with Chilean officials and key players. Each meeting has a description of what was discussed, as well as the individuals we met with and which teams met with them.

| Date | Meeting, Location, and Key Officials | Topics Discussed |
|---|---|---|
| July 19 | ONEMI, Central Alert System (CAT), Santiago [All]<br>---Johaziel Jamett, CAT Manager<br>---Andres Salgado Saint<br>---Anne, Cabinet Chief<br>---Cristian Torres F., Institutional Development Coordinator | ---Need for better interagency coordination<br>---Impacts of earthquake (damage #s, etc)<br>---Importance of practicing operational plans<br>---Communications systems restored quickly, overall |
| July 20 | Chilean Red Cross (Chilean Red Cross) National Leadership, Santiago [EM, E, SE]<br>---Lorenza Donoso, Chilean Red Cross National Director<br>---Yolanda Munoz ,Executive Director<br>---Francisca Ringer and Patricio Acosta, Vice Presidents<br>---Nelson Hernandez, Director of Disaster Risk Management<br>---Jorge Romo and Maricella Huentemilla, Logistics Coordinators<br>---Robinson Talavera, Director of Health<br>---Cristian Ramirez, Water and Sanitation Coordinator<br>---Osvaldo Rodriguez (Not Chilean Red Cross), Real Estate Consultant | ---Chilean Government (GOC) has increased the Chilean Red Cross's post<br>---disaster responsibilities<br>---Country's view of Chilean Red Cross is shifting to see them as disaster responders, instead of only being seen as first aid support and disaster educators<br>---Chilean Red Cross shelters were widely unused. It is culturally traditional to stay in ones' own home.<br>---30,000 spontaneous volunteers came forward to help in aftermath---Chilean Red Cross not equipped to receive this high of a volume of volunteers. |

| Date | Meeting, Location, and Key Officials | Topics Discussed |
|---|---|---|
| July 20 | Chilean Red Cross Volunteer Management, Santiago [V]<br>---Francisco Cruzat, Director of Youth Services<br>---Andres Recabarren, Director of Corporate Social Responsibility<br>---Sylvia Santander, Communications Director | ---Currently no specific process for new volunteer intake, nor any specific set training; content is decided by instructors.<br>---Each Chapter handles volunteer management independently.<br>---No national volunteer database<br>---Older volunteers manage the Chapters; younger volunteers often not well accepted. 97% of volunteers are women. Primary "peace time" roles for Chilean Red Cross include:<br>---Youth<br>---Risk Management<br>---Health<br>---Social Activities/Wellbeing (elderly, poor, immigrants) |
| July 20 | Health Services Meeting, Santiago [HS]<br>---Dr. Celso Bambaren, Pan<br>--- American Health Organization/WHO | ---Overview of the earthquake impact to health and medical services<br>---Loss of approximately 4,000 hospital beds<br>---Use and role of temporary and field hospitals<br>---Relationship and agreements between PAHO and the Chilean central government and regions regarding emergency response<br>---Immediate emergency financial assistance (equivalent to $12 million US) from the UN, PAHO—60% of emergency funds used for recovery of healthcare facilities<br>---Government identified healthcare priorities post quake:<br>Need for field hospitals (17 were erected post quake);<br>Need for electrical generators (PAHO purchased 13 for hospital use);<br>3.) Technical support for hospitals<br>---No official national/government report available on damage to healthcare facilities or infrastructure<br>---Many decisions post quake were political and depended upon the expectations of the local, impacted communities |

**(Continued)**

| Date | Meeting, Location, and Key Officials | Topics Discussed |
|---|---|---|
| July 20 | President's Emergency Committee, Santiago [All]<br>---Cristobal Lira Ibañez, Executive Secretary | ---Official GOC damage numbers<br>---Discussed GOC's response to this event, including long-term recovery plans<br>---General lack of preparedness for this magnitude of an event |
| July 21 | Catholic University of Chile, Santiago [SE]<br>---Rafael Riddell, Professor of Civil Engineering<br>---Ernesto Cruz, Professor of Structural Engineering | ---Single---family dwellings mostly built of confined masonry, some are built of adobe; very few are built of wood frame<br>---Three or four collapses of engineered reinforced concrete buildings occurred during earthquake shaking<br>---Professor Riddell attributed most of the damage to engineered buildings to structural irregularities |
| July 21 | Health Services Meeting, Santiago [HS]<br>---Dr. Vladimir Pizarro, Director of Recovery for Ministry of Health, former Director, Hospital Felix Bulnes, Santiago | ---Overview of evacuation of Felix Bulnes Hospital, Santiago (only hospital in the Santiago area to evacuate entirely)<br>---Entire hospital (275 adult/children patients on floors 2-7) evacuated in 17 minutes<br>---All but #20 ventilator---dependent patients evacuated to parking lot<br>---All evacuees were transferred to other hospitals or discharged within 24-48 hours<br>---Ministry of Health has a centralized system to coordinate hospital bed occupancy |

| Date | Meeting, Location, and Key Officials | Topics Discussed |
|---|---|---|
| | | ---Felix Bulnes had an emergency plan but no centralized plan for evacuation<br>---Ministry of Health reviews hospital plans but has little ability to enforce<br>---Overall, approx 3,100 patients throughout Chile had to be transferred to other hospitals due to earthquake damage<br>---Field hospitals were used primarily for new patients (the timing was such that most were not set up in time to accept transfers)<br>---No reports of death/injury associated with hospital evacuation/transfers |
| July 22 | Maule Region Chilean Red Cross and other key regional players [All]<br>---Cristian Winkler, Chilean Red Cross Regional Exec. Dir.<br>---Marcella Winkler, Chilean Red Cross<br>---Francisco Sans, Civil Engineer (regional government)<br>---Maria Soledad Acuna, Social Assistant (regional government)<br>---Orlando Saavedra, Political Scientist, Government of Talca | -Two options for government---supplied emergency housing: move to 'village' with others displaced by tsunami and earthquake; or have temporary housing on one's own property<br>-Many affected people preferred to remain on their property to protect belongings instead of going to shelters –<br>-16 Mediasaguas Villages in Maule Region housing about 700 families<br>-Towns in coastal parts of Maule Region have tsunami drills every two months<br>-Many cities execute extensive emergency drills 1---2 times yearly, including evacuations to high ground<br>-In the Maule Region, GOC distributed 40k boxes of food which fed 4 people for 4 days in first 2 months<br>-In the Maule Region, Chilean Red Cross distributed 5071 boxes of food which fed 5 people for 15 days in first 2 months |

**(Continued)**

| Date | Meeting, Location, and Key Officials | Topics Discussed |
|---|---|---|
| July 22 | Visited Mediasaguas Village<br>outside of Constitución [All] | -Tour of village with Chilean Red Cross and informal discussions with village lead and residents<br>-Village had toilet and shower facilities, which were divided up amongst families.<br>-Each Mediasagua had electricity, but no heating or cooling systems.<br>-Village had group meeting tent, which was originally built for residents to watch the World Cup<br>-Trucks come from the city with goods, produce and meat for people to buy. |
| July 23 | Cauquenes Government and local Chilean Red Cross [All]<br>---Guillermo Garcia Gonzalez, Governor<br>---Jorge Rojas Luna, Provincial ONEMI director<br>---Hernan Larrain Fernandez, Senator<br>---Ignacio Urrtia Fernandez, District Representative<br>---Girardo Arellano Vallejas, Cauquenes District Emer. Mgr<br>---Cristian Winkler, Chilean Red Cross Regional Exec. Dir.<br>---Marcella Winkler, Chilean Red Cross | ---Historical part of Cauquenes, which was comprised mostly of adobe homes, was destroyed<br>---There were approximately 4200 adobes homes in the District of Cauquenes; the 12 people who died in the district were all killed in adobe houses<br>---Many people ran out of homes during earthquake<br>---In Province of Cauquenes, 64 people died, and 3 are still considered missing |
| July 23 | Cauquenes Hospital [HS]<br>---Hospital Director<br>---Public Affairs Manager | ---Tour of hospital and discussions with Hospital Director and Public Affairs Manager<br>---Tour of field hospital set up to replace portions of Cauquenes Hospital destroyed by the quake |

| Date | Meeting, Location, and Key Officials | Topics Discussed |
|---|---|---|
| July 23 | Regional Bomberos, Cauquenes [All]<br>---Carlos Peñalillio, Community of Chanco Fire Chief<br>---Richard Castillo, 3rd Company Fire Chief in Cauquenes | ---Local inspections put houses in following categories: Completely destroyed, habitable, or inhabitable; completely destroyed, major damage, or minor damage<br>---International help started arriving five days after initial quake<br>---All Chilean Bomberos are volunteers, but most local employers kept paying Bomberos employees during rescue efforts |
| July 24 | Meeting with local Bio Bio region Chilean Red Cross volunteers [All]<br>---Manuel Quezada, Regional Emergency Director<br>---Claudia Betancourt, Chilean Red Cross Paramedic<br>---Gladys Rivero, Chilean Red Cross hazards manager for Bio Bio Region (from Concepción)<br>---Angelica Lagos, Chilean Red Cross Paramedic<br>---Ricardo Rios, Talcahuano Hazards Manager<br>---Marcia Gutierrez, Temuco Chilean Red Cross | ---10---12 days for most basic services to resume in Concepción<br>---Military service is required for most Chileans. Those who do not join the military can be a part of the Civil Defense. Both teach first aid and general preparedness actions<br>---Military presence in streets of Concepcion five days after quake, in effort to quash looting<br>---In Tumbes, over 500 people live in Villages<br>---99 families in Village in Talcahuano<br>---Radio Bio---Bio (emergency broadcast station) was the only station transmitting---bandwidth covers most of Chile; initially, a lot of misinformation may have been disseminated |

**(Continued)**

| Date | Meeting, Location, and Key Officials | Topics Discussed |
|---|---|---|
| July 24 | Regional Bomberos<br>Coordination Center (Regional Emer. Operations Ctr.) [All]<br>---Marcos Oyarzo Espinoza, Province of Talcahuano Bomberos Commander<br>---Dino F. Olivieri Diaz, Regional President of Bio---Bio Bomberos | ---VERY few fires caused by earthquake<br>---People causing fires by using damaged heating units<br>---Regionally, 36 fire stations were completely damaged. In Concepcion, four stations collapsed. In Talcahuano, three engines were destroyed<br>---Some streets in tsunami---ravaged Talcahuano were clogged with 1---1.5 meters of debris<br>---Lack of ability to communicate between emergency responder/service entities<br>---Bomberos offered security to last supermarket in town that had not been looted in exchange for supplies for emergency services personnel |
| July 24 | Tours of Cities of Concepción, Talcahuano, and Tumbes [All] ---Local Chilean Red Cross Representatives<br>---City and Regional Fire Personnel | ---Survey/tour of damaged parts of Concepción, Talcahuano, and Tumbes with Chilean Red Cross and regional fire personnel<br>---Visited Alto Rio and Torre O'Higgins buildings (two of most well---known quake---damaged buildings in the country) in Concepción; 8 dead, 71 rescued in Alto Rio<br>---Saw heavily a damaged fire department building in Talcahuano (it is across street from the ocean and was hit hard by the tsunami) |

| Date | Meeting, Location, and Key Officials | Topics Discussed |
|---|---|---|
| July 24 | Visited Mediasaguas Village outside of Tumbes [All] | ---Tour of village with Chilean Red Cross and informal discussions with village lead and residents, and Mayor of Tumbes<br>---Village had five portable toilets<br>---78 Mediasaguas<br>---Each Mediasagua had electricity, but no heating or cooling systems.<br>---Village had group meeting tent, which was originally built for residents to watch the World Cup<br>---Trucks come from the city with goods, produce and meat for people to buy. |
| July 24 | Former Regional Secretary of Health, Concepción [HS]<br>---Dr. Mara Werner, former Regional Secretary of Health, Concepción | ---No ability to communicate internally or externally immediately post quake<br>---By 0600 able to text message other Directors<br>---All clinics/hospitals "did the best they could to respond to the moment."<br>---Loss of 1/3 of the hospital beds and 36% of operating rooms in the Concepción region<br>---First priority was to assess structural damage in the provinces – but little communication with outlying areas<br>---Outpatient dialysis was a "big problem"<br>---No data regarding hospitalizations or injuries post quake, few records were kept (Dr. Werner is attempting to reconstruct some data)<br>---Mental health issues noted almost immediately, started group therapy sessions in the shelters and communities |

**(Continued)**

| Date | Meeting, Location, and Key Officials | Topics Discussed |
|---|---|---|
| | | ---No known water/food rationing but had to be "very careful" – some workers brought food to hospitals, looting also contributed to food availability in Concepcion area<br>---"Seismic Culture" (supposedly) but few are personally prepared w/food, flashlights<br>---Locals were really affected by the disaster, workers got sick and tired – it was difficult to receive external help since they required a great deal of direction from local medical authorities |
| July 26 | University of Chile, Santiago [SE]<br>---Professor Sergio Barrientos, Ph.D. | ---450---500---km---long section of fault ruptured<br>---Approximately one magnitude 7.75 earthquake or greater every 10 years in Chile (over a 450 year average)<br>---Tsunami hit as tide cycles were approaching neap tide (the lowest low tide), so the largest tides came when high tide was approaching (3---8 hours later in some places)<br>---Received isoseismal map, which shows earthquake intensity for the affected regions |
| July 26 | Ministry of Housing and Urban Development, Santiago [EM, E] ---Vicente Nunez, ONEMI director \---Enrique Ganuza (UN)<br>---Francscio Irarrazabal (Housing Ministry) | ---Design plans were presented for the reconstruction of the cities of Constitución (including its island) and Concepción.<br>---Presentation of the Tsunami action plan was presented. Plans include moving the existing port in Constitución to another location for example |

| Date | Meeting, Location, and Key Officials | Topics Discussed |
| --- | --- | --- |
| July 26 | Red Cross National Headquarters, Santiago [EM, E] ---National Director and Staff | ---A tour was provided to the ARC delegation<br>---Chilean Red Cross hosted a roundtable to present their lessons learned. 'Be prepared for everything and anything. Be prepared to have absolutely nothing to support the recovery efforts—water, technology, electricity, and supplies. Everything you believe can go wrong (and more), will go wrong.'<br>---Suggestion to conduct an exercise of emergency and disaster response with absolutely No technological resources at play. |
| July 27 | Ministry of Health, Santiago<br>[HS]<br>---Gisela Acosta, RN, Chief, Emergency Preparedness and Response, Ministry of Health | ---Weak in preparation, good at response<br>---Discovered many strengths, weaknesses<br>---Communication biggest problem<br>---Many places didn't have radio communication, others had systems but weren't maintained (example: no batteries)<br>---Took 2 days for central gov't to have contact w/Concepción and Talcahuano<br>---Main request post---quake was for structural evaluation of hospitals<br>---Donations began to arrive immediately but weren't always what was needed or requested<br>---Public hospital system was very stressed—few private hospitals in the southern region to accommodate surge/transfers<br>---Not as many injured in quake as expected, few major injuries, mostly minor and treated locally<br>---Finding available ICU beds for those in ICU pre---disaster was an issue<br>---ONEMI had a good plan but there was failure to execute<br>---Assistance from police/fire/power workers to get electricity to impacted hospitals<br>---Current priority:New emergency disaster plan with reinforcement of regional and central plans, emphasis on the importance of communication and safe hospitals |

**(Continued)**

| Date | Meeting, Location, and Key Officials | Topics Discussed |
|---|---|---|
| July 27 | National Bomberos<br>Headquarters, Santiago [EM, E, V, SE]<br>---Miguel Reyes Nuñez, National Bomberos Chief | ---In Concepcion, Bomberos had to fight fires set by looters instead of being able to fully focus on search and rescue efforts<br>---15,000 Bomberos helped during earthquake aftermath<br>---The only paid staff are engine drivers and communication staff (dispatchers)<br>---Country profile of Bomberos:<br>38,000 Bomberos<br>4000 Female Bomberos<br>311 Fire Stations<br>---There are 345 districts throughout Chile, so some stations service more than one district, and some districts have many fire stations |
| July 27 | Ministry of Health, Santiago<br>[HS]<br>---Dr. Jorge Lastra, Director of Medical Response for Ministry of Health in Bio---Bio Region | Dr. Lastra was unable to meet with us. Upon return, there has been e---mail contact and follow---up. The HS Team submitted written questions to Dr. Lastra (translated by Nancy Carrillo into Spanish) and is awaiting a response from him. |

Abbreviations: SE=Science and Engineering Team; V=Volunteer Team; HS=Health and Medical Services Team; EM=Emergency Management Team; E=Executive Team.

# 12.0. Glossary of Terms including Acronyms

AHRQ: Agency for Healthcare Research and Quality *Aldeas*: Temporary housing villages.

ARC: American Red Cross

ARCLA: American Red Cross, Los Angeles

ATC-20: Applied Technology Council--20 (Procedures for Post--earthquake Safety Evaluation of Buildings)

*Bomberos*: Chilean firefighters.

CA DSS: California Department of Social Services

*Comité de Emergencia de la Presidencia*: President's Emergency Committee

FONASA: Fondo Nacional de Salud (National Health Fund; public sector healthcare) GEER: Geo--- engineering Extreme Events Reconnaissance

IFRC: International Federation of Red Cross and Crescent Societies

IOM: Institute of Medicine

ISASPRE: Instituciones de Salud Previsional (Health Insurance Institutions; private sector healthcare insurance)

ONEMI: Officina National de Emergencias, Ministerio de Interior

*Mediasaguas*: Temporary housing shelters built by the Chilean Government, in order to house those displaced by the earthquake and/or tsunami.

MINSAL: Ministerio de Salud de Chile (The Ministry of Health of Chile) MRC: Medical Reserve Corps

NOAA: National Oceanic and Atmospheric Administration

PAHO: Pan-American Health Organization

WHO: World Health Organization

# INDEX

## D



## E

## F



## G



## H

**N**



**O**



**P**

## Q



## R



## S



## T

## U



## V



## W



## Y